THE

COW

A GUIDE TO
DAIRY MANAGEMENT
AND
CATTLE REARING

LONDON WARD LOCK & C°

ROWLANDS' ODONTO

THE COW.

THE COW.

A Guide to

DAIRY-MANAGEMENT AND CATTLE-REARING.

CONTAINING ALL NECESSARY INFORMATION REGARDING

ANIMALS, GRAZING, MILK, BUTTER, & CHEESE.

London:

WARD, LOCK AND CO.,

Warwick House, Salisbury Square, E.C.

PUBLISHERS' PREFACE.

———◦◦◦———

IN the following pages we have a full account of the Management of the Cow and the Keeping of the Dairy, in all its branches, and both on a large and on a small scale. The writer is practically acquainted with his subject, and he has done his utmost to make his work a valuable and standard book of reference.

He has had in view not only to aid those amateurs, resident in the country, or in the outskirts of our large towns, who keep cows for the use of their own establishments, but to be of service also to the professional farmer, who will find in these pages a complete summary of those scientific methods of Dairy Farming in which so great an advance has of late been made.

The publishers believe that in these endeavours the author has fully succeeded, and that this contribution to the literature of the Dairy will find a place on many a book-shelf, and be often consulted to great advantage.

For several of the Illustrations appearing in this volume, the Publishers are indebted to Messrs. BRADFORD & Co., High Holborn; Messrs. ALWAY & SONS, White Lion Street, Pentonville, London; and Messrs. R. HORNSBY & SONS Grantham.

CONTENTS.

———◆◆◆———

LIST OF ILLUSTRATIONS.

———◦◇◦———

FAT SHORT-HORN STEER.

THE COW.

CHAPTER I.

ENTERING ON A DAIRY FARM.

Questions to be Considered before Commencing a Dairy—Comparative Profits from Making Cheese or Butter, Selling Milk, or Grazing—Dairy Operations in mixed Husbandry Farms—Education necessary for the Dairy Farmer—Rent, Capital, &c—Most Suitable Time of Entering—Stocking the Farm.

1. BEFORE STARTING A DAIRY, there are several points which ought to be seriously considered. First, from the situation of the farm, it should be taken into account what it is best to aim at in the way of production—whether milk-selling, butter-making, or cheese-making. This will mainly depend upon facilities for carriage, and the proximity of large towns, upon which often depends the profitable disposal of produce. Next, if in a country situation, where the cows can graze the meadows, the quality of the "feed" should be regarded, for there are many of the smaller kinds of cows that thrive well enough upon rather poor herbage, upon which larger and heavier animals would nearly starve.

In some of the South-Western districts of England, for example, Alderney cows do well upon grass quite unsuitable for the large animals of the old Yorkshire stock, and other similar kinds, which are so much in favour with the London cow-keepers, who provide them with large quantities of food, for the purpose of forcing the milk artificially.

2. ARTIFICIAL FEEDING AND NATIVE PASTURE.—Although much may be said in favour of artificial feeding under certain conditions, yet it must ever be borne in mind that, after all, there is no food which can be compared with that of good natural pasture for milch cows; for not only do they thrive on it, and give a larger quantity of milk, but the flavour of the butter is richer and more delicate, and, in consequence, commands a higher price in the market. It is best to settle these points in the first place, as there is always a drawback in changing one's system, and nothing answers so well as when matters are conducted upon a principle of continuous routine, where everything falls naturally into its place, with system and order.

3. SIZE OF BREED.—If the Breed of Cattle is too large for the quality of the pasture, the return in the shape of produce will be considerably less than it ought to be, with more discriminate management; as the bulk of the food consumed will be absorbed in the office of keeping up the animal's system, instead of producing milk.

On very rich pastures it matters not how large the breed of cattle that is placed on it is, as they can obtain an abundance of food, and the yield will be correspondingly heavy. On the whole, medium-sized animals are found the best for dairy purposes, as they are able to maintain themselves upon pastures of an average quality, and they are less likely to become affected when, from certain causes, the ordinary feed becomes temporarily deficient. By skilful feeding, however, these breaks can be so regulated, and their effects lessened, that no serious inconvenience is to be apprehended on this score, by one who thoroughly understands his business, however inexperienced dairy farmers may suffer upon these occasions.

4. HOUSE ACCOMMODATION. — Cold winds in spring and summer, if cows are exposed, prevent a full flow of milk, and there should be sufficient accommodation for house-feeding, as well in the summer as the winter months. During a hot summer, again, the herbage suffers from drought, and during these times, if protracted, the cows stand a chance of being seriously injured, unless food of a nature calculated to supply the deficiency is given to them; so that house accommodation is a point that should not be overlooked.

5. BUTTER-MAKING.—If in an isolated district, where there is no large demand for milk, it will be found best to produce butter.

Railway communication is now so complete and perfect in all parts of the country, and butter is comparatively such a portable article, that disadvantages of situation can be atoned for to a very considerable degree by the aid of the railway.

6. CHEESE-MAKING is more of a manufacturing business, which, to be carried out successfully, requires a considerable amount of technical knowledge, and also experience, and much previous practice. The English cheese-maker has also to enter into competition with cheese produced in America and on the Continent, where it is made upon a very large scale and by a thorough routine system. It will not be found to answer so well (save in very exceptional cases) as disposing of the produce in the form of butter or milk.

7. MILK PRODUCTION.—But of all ways of disposing of dairy produce profitably, nothing answers so well as to get rid of it at once, in the form of milk, as it comes warm from the cow. When this system, from proximity to a large town, or where there is easy access to rail, can be conveniently adopted, it is unquestionably by far the most advantageous plan which can be followed by the dairy farmer.

Milk has steadily risen in price of late years, and since the Adulteration of Food Act has come into force, and a purer article is supplied to the consumer, its value as a diet has risen considerably in public estimation, as its quality can now be relied on; and no matter how other articles of food may fluctuate, the price of milk steadily remains the same, and its price is relatively higher than that of the manufactured article when converted into butter or cheese. And under the most favourable circumstances for the sale of his produce, the maker of butter can seldom hope to reach the average from each cow which can be made from the sale of new milk, even under ordinarily favourable circumstances, although the price of butter has risen considerably of late years.

8. COMPARATIVE PROFITS.—A prize essay on the profits of grazing, making cheese, and selling milk, written by Mr. W. H. Heywood, was published some years back in the Journal of the Royal Agricultural Society, which showed the profits, in the instance selected, to be much greater from selling milk than from the two other methods, which ranked respectively, milk-selling first, grazing second, and cheese-making last. It is termed cheese and butter making, but the results of cheese manufacture are alone given particulars of, it being assumed that cheese-making and butter-making are equivalent. According to our experience, however, butter-making, where a good market is obtainable for the article, is much more profitable than cheese-making.

The example adduced by Mr. Heywood is a very good one, inasmuch as it is not a comparison between two different farmers, but the results upon the same farm, from different methods practised by the same tenant, who is described as an excellent farmer, and who therefore may be assumed to know the best methods applicable to each course of management. The farm was originally managed as a cheese-farm, up to a certain time, when, in consequence of the advantage of a railway station within a mile of the farm, and twelve miles from the market town, the tenant sold his milk, delivered at the station, at 1s. 10d. per dozen quarts, keeping the management of the farm in other respects precisely as before; the stock and expenses remaining also the same, except that the number of pigs fattened was reduced.

"I will take the case of the cheese-farm, 200 acres, upon which the stock is 50 milk-cows, 50 ewes (which, with their lambs, are fed off fat), 5 horses, 30 pigs, reared up and fattened, and 12 to 15 young horned cattle, consisting of calves, yearlings, and two-year-olds. The farm is self-supplying as regards all food for stock, having sufficient land under plough, viz., 45 acres in 15 acre shifts—ley-oats, turnips, and wheat—to grow the oats, turnips, and straw required, in addition to the old meadow hay.

"*The financial results* of this farm have been as follows:—

PRODUCE.		£	s.	d.
9 tons 7 cwt. 2 qrs. cheese, at 80s. per cwt.		750	0	0
70 lambs, at 27s. 6d.		96	5	0
Profit on 60 ewes and wool, at 15s.		37	10	0
15 acres wheat, at £12		180	0	0
Profit on 30 pigs, at £5		150	0	0
		1,213	15	0

EXPENSES.		£	s.	d.
Rent, 200 acres at 40s.		400	0	0
Tithes, 3s. per acre; rates, 2s. 6d. on assessment		58	15	0
Wages—5 men at £40		200	0	0
2 lads at £20		40	0	0
Extra men		26	0	0
Harvesting		30	0	0
Tradesmen's bills, £52 10s.; grass seeds, £22 10s.; other seeds, £20		95	0	0
Paid on Improvement Account, including Draining £40, Boring £60, and Repairs £25		125	0	0
Contingent Expenses		50	0	0
		1,024	15	0
Profit		189	0	0

"*The result under the system of milk-selling* is as follows, more milk having been produced per cow in consequence of the supply having been kept up throughout the year by exchange of cows and artificial feeding:—

PRODUCE.		£	s.	d.
Milk of 50 cows, at 1s. 10d. per dozen quarts		1,065	0	0
70 lambs, at 27s. 6d.		96	5	0
Profit on 50 ewes and wool, at 15s.		37	10	0
15 acres of wheat, at £12		180	0	0
Profit on 10 pigs, at £5		50	0	0
		1,428	15	0

EXPENSES.		£	s.	d.
As per statement in Cheese-making Account		1,024	15	0
Add cost of exchanging cows to keep up supply of milk at certain seasons		100	0	0
		1,124	15	0
Profit		304	0	0

"*On the grazing-farm* referred to the stock is 60 cows, 100 ewes (whose lambs are fed off fat), 4 horses. The result is as follows:—

PRODUCE.		£	s.	d.
Profit on 60 cows, at £12		720	0	0
140 lambs, at 27s. 6d.		192	10	0
Profit on 100 ewes and wool, at 15s.		75	0	0
15 acres of wheat, at £12		180	0	0
		1,167	10	0

EXPENSES.				£	s.	d.
Rent, 200 acres, at 40s.				400	0	0
Tithes, £15; rates, £43 15s.				58	15	0
Wages—4 men at £40...	160	0	0			
1 man at £20	20	0	0			
Extra man	13	0	0			
Harvesting	20	0	0			
				213	0	0
Tradesmen's bills, £32 10s.; grass seeds, £22 10s.; other seeds, £20				75	0	0
Paid on account of Improvements, including Draining £40, Boring £60, and Repairs £25				125	0	0
Paid for oil-cake				50	0	0
Contingent Expenses				30	0	0
				951	15	0
Profit				215	15	0

"The three systems will, therefore, stand as follows:—

	Receipts.	Expenses.	Profits.
Cheese or Butter-making...	£1,213 15 0	£1,024 15 0	£189 0 0
Grazing	1,167 10 0	951 15 0	215 15 0
Milk-selling...	1,428 15 0	1,124 15 0	304 0 0

"*It thus appears* that the experience of this district (North Cheshire) is decidedly in favour of milk-selling; but before coming to a definite conclusion on the subject, the strain put upon the land by the two systems—milk-producing and fattening—has to be taken into account.

"I feel that the grazing account may require some little explanation to some whose experience may be somewhat different. The profit of £12 per head on the cows may be thought excessive. I can, however, but state that such is the annual average profit realized by a number of graziers in this immediate neighbourhood, who buy in lean but healthy shorthorns, at an average of £10 to £12 per head, in the first two months of the year. They then freshen them on straw, turnips, and a little cake, putting them out a little each day—weather permitting—until spring, by which time they have fairly begun to grow; and when a flush of grass comes they do not, like cows newly bought, lose time in making a start. They are then grazed through the summer, tied up in October to turnips, ground oats, oil-cake, and straw, and sold from the middle of December to the middle of January at £22 to £24 per head. The extent of land may also seem small for the number of beasts and sheep kept; but this is accounted for by the circumstance that all the grass land is available for pasture, only a small quantity being required for the horses. Again, the practice is to break up a fresh turf-field every year for ley-oats, to be succeeded by turnips, which, aided by the moist climate of the district, is always a very heavy crop, averaging from 33 to 38 tons per statute acre: hence the large amount of winter-keep from so small an extent of arable land.

"The item of £50 for cake may also appear small, but I may state that cake is not used as the chief article for fattening beasts, but rather as conducive to their health, and as an aid to the corn and turnips, which are mainly relied upon for fattening them. The sheep and lambs get no cake.

"I may also further state that of the 60 cows grazed, not more than 50 are tied up in the autumn, as the remainder either go out from grass or as calvers, of which there are always a few, and which pay equally well, regard being paid at the time of purchasing that they are all right in their milking organs.

"But I should hardly do justice to the merits of this system of grazing by simply giving the practical results in my own neighbourhood, and comparing them financially with those of cheese or butter making, and milk-selling. Grazing has collateral advantages in many forms that do not show themselves in such a comparison, but which assume so large an amount in the aggregate, that, though milk-selling excels it in direct profit by, say, £88 5s. per annum on a farm of 200 acres, I yet consider that, in the main, grazing is the preferable system, as I will endeavour to show.

"In the first place, I consider that the apparent margin in favour of milk-selling may fairly be reduced somewhat, on account of the extra risks attending the system, from the more general tendency to delicacy and sickness, of milking, as compared with fattening, cows. Again, we must not overlook the risk of making bad debts with the milk-dealers; who, as a body in the large towns, are not the best of payers. In saying this I do but speak the experience of milk-producers. Again, under the system of grazing, the farm will regularly increase in fertility, as a much greater portion of the nutriment, either extracted from the ground or artificially supplied, is then returned to it again by the animal, than under the system either of cheese-making or milk-selling. If, then, we suppose a tenant to have a lease for, say, twenty-one years, at a fixed rent, the progressive improvement of his farm under grazing will yearly increase his crops of beef, mutton, and corn; and with improved condition of land comes decrease of expense in cultivation; and thus his profit will yearly go on increasing, the ultimate result being most beneficial alike to himself and his landlord.

"*As regards the labour* attending the practice of these systems of farming, that of grazing has a decided advantage over the others, not only in out-door labour, as shown in accounts of expenditure, but also in the labour and responsibility saved in-doors, since the care and management of milk, in any way, entails much of both, and requires an amount of skill that has often to be remunerated at a very high rate.

"*One of the best indications* of the progressive improvement attendant on this system of grazing is obtained by one simply observing the very great difference in the quality of the dung-heaps collected under the respective systems, the comparatively cold, aqueous appearance of that produced from milking-stock contrasting remarkably with the fermenting, oily nature of that collected from fattening-beasts. The effect of this difference upon the farm must be obvious to anyone. In fact, I have myself watched its progressive effect under good management with extreme satisfaction, seeing the ordinary condition of the farm rise gradually to that of high cultivation; the weeds disappearing as the crops became stronger, and the land being more easily worked as it became more disintegrated by the more luxuriant growth of the herbage upon it. Here I cannot but state the particular attention paid by the farmers of this district to the mode of seeding down their pastures, which, coupled with the clean fallow, or green crop, is undoubtedly, after draining, the foundation of all good farming, and the secret of success in the cases now under my notice. By attention to this particular, a sod is obtained by the aid of bones which, after a few years' growth, is equal to that produced in the ordinary way by twenty years' ley ; and experience shows me that a good sod that breaks up oily and mellow, through the action of the fibres of luxuriant herbage, conduces more to a good and inexpensive course of crops than any manure that can possibly be applied artificially, to say nothing of the economy of restricting the need for such manures ; for, after all, artificial manures are but a defective substitute for the elements as naturally combined in a virgin soil.

" Holding these views, and considering the present scarcity and consequent high price of beef and mutton, I cannot commend too strongly a system so conducive to the mutual advantage of both tenant and landlord as that of grazing."

There can be no two opinions as to the remunerative nature of grazing and stock-keeping, and the great importance of having more stock on a farm than is usually kept has been long urged upon the notice of agriculturists by different writers ; but it is possible that Mr. Heywood, in drawing his comparisons between milk-producing and grazing, has been guided to his conclusions by the methods which are followed mostly by farmers in this country, who feed their milch cows only upon grass in summer, and the cheapest substitutes which they can procure in winter, when he points to the great difference in the quality of the manure ; for, of course, with cake, pea-meal, and other higher feeding, which, it is contended, it pays to give to milking-cows for the sake of a greater yield of milk, the advantage in extra richness of manure is disposed of.

9. MARKETS FOR MILK.—There is always a ready market for any amount of milk in London, and the case is much the same in the neighbourhood of our principal towns and cities.

10. DAIRY OPERATIONS ON MIXED HUSBANDRY FARMS.—The dairy, and dairy produce may be the chief aim, but there are many subsidiary items which all come in to swell the profits of the business, and these deserve the most careful consideration, though they are very often too much neglected.

Pigs, although, strictly speaking, having no connection with the dairy, can be fattened on skimmed milk, and calves cheaply reared on the same ; which in time grow into valuable stock, that can be made to possess the additional recommendation of being bred to one's actual requirements by judicious management.

The best roots also may be sold which can be spared at times ; such as carrots, parsnips, &c., as well as cabbages and similar pro-

B

ducts, which can all be consumed if there does not happen to be a market; and the same will apply to hay, or any produce, if the pasture-land happen to be in excess of the dairy farmer's own requirements.

11. EDUCATION NECESSARY FOR THE DAIRY FARMER.— There are many points in connection with dairy management in which the inexperienced require to be educated, so to speak, in order that definite results may be obtained, and the business not given up to haphazard, or anything left to chance.

The necessity of this is shown in the example of many struggling men, who work hard, and live frugal lives, and yet can barely obtain more than a decent subsistence, instead of laying by money, as they ought to do; while the amateur in most instances, instead of making money by his dairy farm, often loses a good round sum annually.

We have to fear the competition from butter-makers of Normandy and Holland chiefly, but there would be less cause for apprehension if English dairy-farmers paid more attention to details than they are generally in the habit of doing. There is a good market at our own doors for all kinds of produce, which is the first essential in any commercial undertaking, and without which many praiseworthy efforts would be thrown away; so that there is no fear but that the English dairy-farmer will be amply repaid for all and any effort he may make to improve the routine of his production, for which, it must be confessed, there is great need in many parts of the country. In others, the necessary precautions for ensuring the health of stock are often neglected, and advantage is not taken of the best methods of adding to their food judiciously, by giving artificial aids which stimulate an increased production. In the rich county of Gloucester, almost the entire food of the cows is grass in summer, and hay in winter; and though doubtless this is their natural food, it might, at times, be most usefully supplemented. Again, scarcely any shelter is provided for the milch cows all the year round, "according to the custom of the county," where the old-fashioned system of dairying prevails, which, it must be acknowledged, is a very bad one in the case of dairy stock, whose yield of milk is increased, and whose health and condition is greatly improved when carefully housed, or partially housed. Various examples are quoted in the following pages of experiments and results arrived at by different persons, in different places, and in those cases where the prices of stock and produce instanced are considerably lower than those which rule at the present time, an approximate allowance must be made, and they must be read with relative application to present rates, which are much higher for nearly all dairy produce (cheese excepted) than they formerly were; and they are therefore, necessarily, proportionately remunerative.

12. RENT, CAPITAL, &c.—Arable land, as a rule, maintains a much steadier and more equable value in the shape of rent than meadow land, which varies exceedingly according to situation, sometimes pasture land in the neighbourhood of large towns commanding high prices for the sake of the accommodation it affords, many large butchers, and others, being willing, at times, to give as much as

£6 per acre for the convenience of turning beasts and sheep into it; to be ready for slaughtering when wanted; besides being needed for; many other purposes, which it is unnecessary to specify.

It is difficult, therefore, to instance, with any degree of certainty,' the amount of rent which should be paid for land where dairy and mixed husbandry farming is carried on; and we should not like to commit ourselves to any definite statement on this head, but a practical man of our acquaintance informs us that he could always make a profit out of grass land for which he paid any sum under £3 per acre.

Of course, in such a general statement, very poor land must be excepted for which a rent of £3. might be demanded, but which

DEVON STEER.

could only be called pasture land by courtesy; so that rent must be regarded in conjunction with *quality*. Very rich pasture land would be much cheaper to the dairy-farmer at a comparatively high price, than very poor, thin, unproductive meadows, that might be rented for very little money.

13. **AN APPROXIMATE ACCOUNT.**—We give a short approximate account of the amount of capital required, rent, &c., in the case of a partly arable dairy farm, in an agricultural district where rents are low, of small size, furnishing work enough for one pair of horses, of 120 acres in extent, 60 of which are pasture—30 acres being mown every year—and the remainder arable, cultivated on the six-course rotation, *i.e.*, 1. Wheat; 2. Beans; 3. Wheat; 4. Clover; 5. Oats; 6. Mangold, Carrots, Turnips, or other roots; the rent of which is 32s. per acre.

The winter stock of food may be put down at 250 tons of roots from 10 acres of arable land, and about 30 tons of hay; and the summer stock at about 400 tons of green food from the clover and grass of the artificial and natural meadows. This would provide for the same amount of stock winter and summer, and support a dairy of 25 cows, eating nearly 2 cwt. per diem each, thus provided.

	£ s. d.	£ s. d.
To purchase 25 cows, therefore, at £20 each, will take	500 0 0	
To purchase 10 store pigs	15 0 0	
		515 0 0
The out-going tenant will probably require to be paid for the cultivation of the young crops. £3 10s. per acre for 60 acres ...	... ...	210 0 0
The labour on the farm, including cost of horses and their keep, may be put down at £4 per acre for 60 acres of arable ...	240 0 0	
And 60 acres of pasture land, at 15s. per acre	45 0 0	
		285 0 0
The implements for the necessary use of the farm are generally put down at 30s. per acre. 60 acres at 30s.	90 0 0	
Dairy utensils	20 0 0	
		110 0 0
Rent and taxes, put down at the low sum of 32s. per acre		192 0 0

The following being a recapitulation of the whole:—

Stock	515 0 0	
Out-going Tenant	210 0 0	
Labour	285 0 0	
Implements	110 0 0	
Rent and Taxes...	192 0 0	
	1,312 0 0	

The above is considered a sufficient amount of capital to work a farm of the dimensions we have specified. There is no doubt but that many dairy-farmers, commencing in a humble way, have succeeded upon much less; but when a painstaking man works early and late, his exertions take higher rank than those of paid labour, for the magic of ownership accounts for many otherwise astonishing results.

But in the regular way an insufficient capital in farming means that a man will be continually behindhand in his operations, unable to take advantage of favourable seasons, or to guard against those which are likely to be adverse: and, as a rule, the more capital that is employed (so that it is judiciously made use of) the better is the result as far as profits are concerned.

14. MOST SUITABLE TIME OF ENTERING.—The autumn is the usual time of entry in England, leases generally terminating at the Michaelmas quarter, the reasons seeming to be in favour of commencing a fresh tenancy in the autumn—first, because the corn crop has been removed from the ground, and second, because the benefit of the summer's grazing has been enjoyed by the out-going tenant; but in the north of England the 12th of May is generally fixed as the time of removal of an out-going tenant, and in the south of Scotland at about, or upon, the 15th of May, or Whit-Sunday.

There is a special recommendation in the latter arrangement to the dairy-farmer, as the cattle are then changed from the folds to the fields, and there are greater advantages on the side of a spring than of an autumn term of entering upon a farm.

The out-going tenant having threshed all his corn of the last crop, and sown the seed of the crop he is entitled to of the following harvest, called the away-going crop in the North, and having consumed all his turnips and hay, the in-coming tenant takes possession at a time when his stock can generally depend upon grass, and he has the making of hay, and the working of land for turnips and mangold, or any special crops he may require in his own management.

In the North, where this system is most commonly practised, he does not get possession of that portion of the land under the away-going crop, which it is customary for the out-going tenant to reap and thresh. This double occupation is, however, very inconvenient, and often leads to disputes and misunderstandings between the out-going and in-coming tenants; it is, therefore, much better avoided, which is generally done by means of valuation.

15. STOCKING THE FARM.—As scarcely any two farms are alike, each possessing its own particular capabilities, or otherwise, one of the most important considerations, to be carefully entertained, is to stock it to the best advantage. We are, of course, assuming the proper feeding of cows to be the principal object aimed at, so as to ensure the largest possible amount of dairy produce; but unless there is ample space for raising plenty of roots, and growing green crops, it will not be wise to be encumbered with too large an amount of live-stock in the shape of sheep, or other than milch cattle, the proper keeping of which may, at times, become a source of embarrassment. With respect also to the quality of the herbage on the permanent pastures, as we have pointed out before, this should determine, to a great extent, the breed of cows that are to be kept the various points of which we shall afterwards specify.

16. THE REARING OF CALVES, again, can be made very profitable, if the business is set about in a right manner, though this branch is often neglected to a surprising extent by most dairy-farmers.

17. PIGS are another paying item, when properly managed, but if too many are kept, and a large quantity of food has to be bought for them, there are but few persons who can make them answer, though

this can be done readily enough by those who understand the best methods for bringing about this desirable result.

Although Arable Farming often needs a large amount of capital to be carried on successfully, and the variations of seasons and risks of bad ones have to be taken into account, while profits can never be large, there are comparatively few obstacles to the breeding of stock, or dairy-farming, the profits on which are much larger than when the cultivation aimed at is the production of cereals only.

LEMORISEIN BREED.

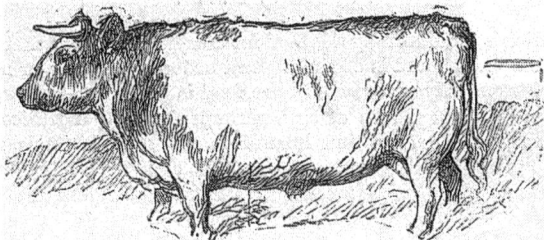

FAT CROSS-BRED STEER.

CHAPTER II.

SELECTION OF CATTLE FOR DAIRY PURPOSES.

Ayrshire Breed—Alderneys—Short-horns—Long-horns—Brittany Cows—West Highland Breed—The Galloway—Herefords—The Suffolk Dun—Irish Cows—Qualities Common to all Good Milking Animals—The Guenon Theory—Choice of Animals, and Length of Time to Keep Them—Best Age for Milch Cows—Management of Stock.

18. PURCHASING STOCK.—If the farm is entered upon at Michaelmas, or Martinmas (November), it is best to purchase stock at the time when they have received the bull, so as to accustom them to their new *habitat* for some months before the time of calving, as stock seldom thrives well immediately after being removed to a new farm, this being very clearly apparent if cows are shifted about the beginning of summer, when they are in full milk, the supply of which is easily affected by a difference of water, or pasture.

In stocking a dairy-farm, the common practice is to buy of the breeders, in April, a number of heifers which have completed their second year. Supposing the tenancy to commence at Lady-day, or Whitsuntide, they are put on the grass, and the bull admitted to them about the middle of July. They are allowed to graze during summer and autumn, and housed in the winter, about the beginning of November.

For milk dairies, cows which give an abundance of milk are wanted —no matter what its quality, which is of secondary consequence to quantity. For butter and cheese making, on the other hand, the richness of the milk is a very important consideration, and the dairy-farmer should thoroughly satisfy himself, whether the cows which give the most milk are actually the most valuable to him.

19. THE AYRSHIRE BREED.—In its native county every pains have been bestowed to develop the milking powers of the Ayrshire

cow, which is so admirably adapted for dairying purposes that it cannot be surpassed, and is of the highest order. On poor or medium soils it is especially useful, where the food is not over good; and they turn out the best payers of any, perhaps, in those localities where the herbage is anything but luxuriant. As stock animals, for the purpose of the grazier, they are not well adapted, as the bullocks are difficult to fatten, and come light to the scale, and the beef is

SWEDISH COW.

coarse in quality. A cross, however, between an Ayrshire cow and a Short-horn bull will produce a good animal, and these are held in favour by the graziers of the west of Scotland.

20. **ALDERNEYS.**—The Alderney cow resembles very much in appearance the Ayrshire, and, from the superior quality of the cream and butter, is held in high estimation by private families. There is. however, a very opposite difference in the excellence of the two breeds, that of the Ayrshire consisting in the abundance of the milk yielded, while that of the Alderney consists in its richness of quality,

For grazing purposes they are not at all suited; their one good point being confined exclusively to the richness of the milk yielded, the quantity being but small; and it will not be found expedient even to feed a steer for the butcher that may happen to be raised; for although they are known to " cut up " better than the butcher

NORFOLK OR SUFFOLK POLLED BREED.

himself imagines they will do, perhaps, at the time of purchase, the *seller* does not get the advantage from this consideration.

21. **SHORT-HORNS.**—The Short-horn breed are capital milkers

SUSSEX BREED.

when there is an abundant supply of food for them, and they are universal favourites, especially in those districts where the usual average of arable husbandry is carried on. Their aptitude to fatten, which is such a valuable qualification in the eye of the grazier, is, however, somewhat objectionable to the dairy-farmer, who wants milk and not meat. Their quiet temper and symmetrical forms, combined with their rich colours, cause them to be universal favourites. Their prevailing colour, and that which is liked best, is black, with deep orange on the naked parts.

22. LONG-HORNS.—At one time Long-horns were the prevailing stock in most of the midland counties of England, but they have gradually given way, year by year, in favour of Short-horns, even in those districts where they have been the prevailing breed from time immemorial.

23. BRITTANY COWS.—This is a small breed of animals which is sometimes kept as fancy stock by gentlemen, but they do not answer the purpose of the dairy-farmer. The presence, in fact, of these smaller-sized beasts in a neighbourhood is often attended with a certain degree of inconvenience when the calves are saved, as stock

FAT SHORT-HORN COW.

is apt to get deteriorated in time by their admixture with the prevailing breed of a district. They are, in consequence, only interesting as a variety to those who are curious in such matters, but are not worth the attention of the dairy-farmer.

24. WEST HIGHLAND BREED.—This is the prevailing breed in the Highlands of Scotland, especially in the larger Hebrides. It is admirably adapted for districts where the pasturage is coarse, and will not only thrive, but will ultimately put on plenty of flesh, where the more tender Short-horn could scarcely exist. They are also known by the name of Kyloes. The cows yield very rich milk, but give only a small quantity of it, and, besides, have a tendency to soon get dry, which causes them not to be desirable for dairy purposes, except in those rugged situations where the " keep " is not sufficiently good for the better kinds of milch cows, which would not answer, nor succeed in a cold, humid climate, upon coarse herbage.

25. THE GALLOWAY.—The Galloway is a similar breed to the above, only without horns, possessing a larger frame than the West Highlanders. They are also of a more quiet and inoffensive disposition, which admits of a greater number being kept together in the same enclosure than of any other breed. The Galloway is more adapted to a lower range of pasture, and more sheltered plains, than the preceding, but even in their native district they have been supplanted for dairy purposes by the Ayrshire.

FAT HEREFORD HEIFER.

26. HEREFORDS.—In their native district, where they are most commonly to be met with, there are many who maintain that the breed is equal to the Short-horn, and their merits are doubtless great as a grazing breed, suitable for fertile soils; but although both the Hereford and the Devon are fine animals, they do not answer as dairy stock, however well they may turn out as beasts destined for the butcher.

27. THE SUFFOLK DUN.—This breed appears to be indigenous to Suffolk, and possesses an undoubted capacity of yielding a large quantity of milk in proportion to the food they require, the dairy

produce of the county having enjoyed a high reputation for a great length of time. They are ungainly in their form, being without horns, and resembling somewhat the polled breeds of Scotland. The prevailing colour used to be a mouse dun, from which they have taken their name; but this hue has latterly changed to a pale red. For the combined purposes of the dairy and the fattening-stall, the Short-horn is, even in Suffolk, fast taking the place of the original stock.

28. IRISH COWS.—Some of the Irish cows turn out very well,

KERRY BULL.

especially the small Kerry, both as good milkers and also for getting into good condition when the time comes round for disposing of them, when they are wanted to put on flesh; but there are a great many very indifferent ones amongst the ordinary run of Irish cows —the Kerry being the best.

29. QUALITIES COMMON TO ALL GOOD MILKING ANIMALS. —The quality common to all good milking animals consists in the tendency to produce milk instead of laying on flesh. On this account the Short-horn is not so good a milker as many others, however desirable the breed may be on other points. A Short-horn cow will give as much milk as an Ayrshire, but consumes a good deal more food, and is, therefore, a much less profitable animal to keep.

FAT DURHAM COW.

SUFFOLK BREED (FAT).

For *mixed Arable and Dairy Farming*, the breeds which have been found to answer best in Scotland, where young stock are reared, are the Ayrshire, Fifeshire, and Angus breeds, or a first cross of either with a Short-horn; and in England a cross also between a Short-horn and an animal of inferior breed, as respects meat-making qualifications, but one which gives a large supply of milk for the food consumed, is preferred, some one or other of the breeds we have indicated as being good milkers.

KERRY HEIFER.

The highest bred cows, it must be remembered, are not the best milkers, and often the ugliest cow in the herd yields the most milk. Good milkers invariably show very angular outlines; for it cannot be expected the cow should be yielding a large quantity of milk and putting plenty of flesh upon her bones at the same time.

There are no reliable signs by which one can be guided in purchasing a cow beyond the animal's good points which present themselves for inspection, and its eneral likely appearance, combined with a knowledge of her breed.

Mr. Stevens in his book on Farming says :—

"As the colour of Short-horns is a prominent characteristic of them, I may mention that roan is a handsome colour, and is, I believe, the general favourite now, the fancy for colour having gone from the red to the white, and is now settled on the roan. Dark red usually indicates hardness of constitution, richness of milk, and disposition to fatten; light red indicates a large quantity of thin milk and little disposition to fatten; but the red in either case is seldom entire, being generally relieved with white on some part of the sides and belly. White was considered indicative of delicacy of constitution, and to get quit of it and, at the same time, avoid the dulness of red, the roan was encouraged and now prevails. The white shows the symptoms sooner than any of the other colours of breeding in-and-in. A single *black* hair on the body, and particularly on the nose, or the slightest blue or black spot upon the flesh-coloured skin upon the nose, or around the eyes, or the least streak of *black* on the tips of the horns at once proclaim that a Short-horn sporting either one or more of these impurities is of mixed blood, notwithstanding all attestation to the contrary."

30. **THE GUENON THEORY.**—A theory which has lately attracted a good deal of attention, called the "Guenon Escutcheon Theory," after the name of the Frenchman who originated it, has been pronounced all "moonshine" by many experienced dairy-farmers. It is based upon the development of the slight fringe above

FAT SHORT-HORNED HEIFER.

the udder of the cow, where the hair points upwards and downwards, which Guenon calls the cow's escutcheon: the longer and wider this is, the more probability of the cow's being a good milker, it is said. No doubt to an experienced eye the general appearance of a cow's udder would present signs by which the judgment would be very materially assisted; but this accidental development of the hair of the cow is pronounced too fanciful to be relied on, beyond, in a certain degree, following the outline and development of the udder, which should be looked at, and not the hair upon it.

In London Dairies a good many interesting particulars were collected together a few years ago respecting the London milk trade, reported to the Society of Arts, and published in their journal of Dec. 15th, 1865. Since that time certain changes have taken place in the trade, in which the working of the Adulteration of Food Act has had some principal share; but there are many particulars relating to the management of a London dairy from which the owners of country ones can take some valuable hints, especially where they relate to such particulars as that of speedily getting rid of unprofitable cows, and the thoroughly systematic manner with which the entire management is conducted, that offers a striking contrast to the careless way that many country dairies are managed, and are yet expected to pay a handsome profit. It is only by the carrying out of strictly business principles that this, or any other calling, can be made to answer in the best degree, and some of the most salient points we shall briefly mention.

In the Selection of Cows London milkmen are guided both by the current produce which the cow yields, and her prospective selling value when they have done with her. Some cows which are tolerable milkers are yet very bad cows for the butcher. To give a striking example, all Alderney and similar breeds come under this category, although they are known to turn out better eventually than their outside appearance commonly gives indications of.

31. **BEST AGE FOR MILCH COWS.**—Thoroughly experienced dairymen do not, as a rule, like very young cows, because their milk is not then at its full yield; nor should a cow be a very old one, because there is some difficulty in fattening her.

The general practice is to buy cows which have had from three to five calves and to keep milking them till they give no more than six quarts of milk per diem. When her milk begins to decrease with the good food that is usually given to a milking-cow, it will be found that she puts on flesh, and is on the road for being in much better condition. Three or four pounds of oil-cake are then given in addition to the ordinary food, and, at the present prices given for meat, she will probably fetch, within a pound or two, as much as was given for her, if bought tolerably cheap; but this, of course, all depends upon circumstances. If second-class beef is low in the market at the time, there may be a greater loss upon her, but this loss will not be a very serious one in the case of the average of a good cow that has been kept for several seasons, speaking generally. It is this consideration which causes the proprietors of many of the first-class cow-houses in London to purchase large-framed animals, wide and straight-backed, deep-bodied, short-horned cows, which display an ability to carry meat, as well as yield milk; though they often cost a good deal of money. These are kept as long as they are profitable, and sold off when their milk decreases, according to circumstances.

Other kinds of cows do not fetch such high prices at the beginning, such as Irish and various foreign cows, that are often found to be very good milkers, and these are disposed of in the same way, whenever it is thought desirable, without any attempt being made to fatten them.

32. **MANAGEMENT OF STOCK.**—In the management of stock, judicious crossing is a main point not to be overlooked; and a skilful dairy-farmer, in the course of a few years, has it in his power to raise an inferior herd of cows into a very superior one, by the ex-

ercise of care and attention; and that without any great outlay in pedigree cattle, for which fancy prices are asked. This can be easily managed if he rears a young bull-calf whenever he wants one. If he has not the particular breed on hand he is desirou. of having, there is no great difficulty in procuring a calf of the description he requires.

A course of selection and rejection should be constantly going on, with the view of maintaining the herd in the highest effective condition; and as the cows get old they should be replaced by younger animals. There is, necessarily, always a natural reluctance to part with a good cow, although she may be a little aged; but a cow should never be kept after she is eight or nine years old. The constitution of cows differs, like that of human beings, and other animals; but when the cow has attained a certain age, her milk is liable to fall off all at once, and one who farms to obtain a livelihood for himself and family, has to look for profit, and, unfortunately, cannot afford to entertain those kindly feelings of attachment for the dumb animals which serve them, which gentlemen or private families may indulge in.

EXTRA STOCK SHORT-HORN.

CHAPTER III.

FEEDING.

Summer Feeding—Stall-feeding by the Peasants of the Lower Moselle—Winter Feeding—Steamed Food for Cows—Methods of Feeding followed by London Cowkeepers—Different Examples of Feeding and Management—Shortening the Cow's supply of Food before Calving—Feeding Cows for Milk or Butter—A course of Good Feeding highly Remunerative—Water.

33. SUMMER FEEDING.—Instead of allowing a limited number of cows to trample down a large area of the growing grass, as is often seen, if one little field is kept for them, into which they can be turned for air and exercise at certain times, it will be found most profitable to resort to stall-feeding while the crops are growing. A large amount of extra food can be procured for them without any great cost, by economical contrivances. For example, the long grass which grows near the hedges in the fields laid down for hay, can be cut, say, six feet round the field, and when it is cut it will grow in length to equal the remaining portion by the time the whole is mowed. This kind of grass is not seed-bearing, and is somewhat rank, but it is, at all events, good green food, and instead of a considerable portion of the grass being comparatively wasted, a large area can be saved to produce hay.

"*What would be thought*," says Sydney Smith, "if we walked all over our bread and butter?" And it will be found more economical to mow as much grass daily as is wanted, even if there is no other convenient food to resort to, rather than have a meadow spoiled

FAT DEVON HEIFER.

FAT DURHAM OX.

entirely. That portion which has been mown will be growing again, and will furnish some nice "feed" when the other comes to be cut.

With some people, however, stall-feeding in the summer-time would be reckoned out of the question, and it is considered good practice to pasture the cows about ten hours daily, upon one or two-year old clover and rye-grass lea, two statute acres being allowed to each animal. During the three summer months, the grass is generally abundant, and the cows are kept in the pastures from 5 a.m. to 8 p.m. each day, and get little else beside, but when the grass begins to get hard, and there is a smaller supply of it, a liberal allowance of clover and vetches is given in the house at night.

When the weather is very hot, and the flies are troublesome, it will be found a good plan to keep the cows in the house during the day, and feed them upon clover, turning them out in the pastures during the cool of the mornings and evenings.

Fields adjacent to the house are to be preferred for grazing milch cows, as the fatigue and annoyance consequent on driving them any considerable distance both lessens the quantity of milk, and deteriorates the quality of the produce made from it, whether it be butter or cheese.

It will be found also a good plan to change the cows from one field to another, as regularly as possible, and have one or two fields shut up, so as to allow the grass therein to grow, and freshen, while the others are being eaten down. By this means the cows will get fresh, clean pasture every ten days or fortnight throughout the summer, which is a very important point, both as regards the quantity and quality of their produce. Where a cow is fed entirely upon grass in the summer, one-and-a-half acres is calculated to be required.

34. STALL-FEEDING BY THE PEASANTRY OF THE LOWER MOSELLE.—Although the method alluded to before is not by any means held up for general imitation, yet the system described by Schwerz, of economically feeding cattle in the district of the Lower Moselle by the poor peasantry, gives a lively idea of what can be effected by economical methods of stall-feeding where the greatest difficulty exists in procuring a sufficient supply of food for the animals. He says:—

"Stall-feeding is general in the Moselle district. In the autumn alone is there some pasturage on the stubbles, and when the after-grass is cut, the meadows are grazed for a couple of hours daily. It is curious to see how the quantity of cattle are fed which are kept on the numerous little parcels of land.

"In the spring the women and children range the fields, cut the young thistles and nettles, dig up the roots of the couch grass, collect weeds of all kinds, and strive to turn them to account. What is thus scraped together is well washed, mixed with cut straw and chaff, and, after boiling water has been poured over the whole, it is given to the cattle. A little later, when the weeds grow stronger, they are given, unmixed, as fodder. The lucerne comes at length to help, and then the clover, which lasts until the autumn, when cabbage-leaves and turnips are to be had. When these are scarce, potato-haulm is taken to help, until the stubble turnips are fit. In winter, cut straw is mixed with the turnips, and warm feeding begins. In the morning a mash of chaff, rape leaves, pea pods, or cut straw, with bruised turnips, potatoes, or oil-cake, boiled up together. Then barley or wheat straw follows this meal, which is repeated at noon and in the evening. In the middle of the day clover or meadow hay is occasionally given to the cattle.

"In larger farms, where ten or fifteen cows are kept, this kind of mash is only given twice a day. The poor farmer is obliged to be more economical, and must

TURNIP CUTTER AND SLICER.

occasionally try to make good the quantity that he cannot bring together. Even in summer he prepares a soup of this kind for his beasts, but then adds clover, thistles, convolvulus bind, and other weeds, to the mixture. A portion of oil-cake is added while it is hot.

"Turnips carefully preserved, mangel-wurzel, turnip-cabbage, potatoes, and swedes play their part in the spring and winter fodder."

Accustomed to the rude abundance, and often waste, on a farm, the English labourer is sometimes inclined not only to view such economical expedients with contempt, but extend it to the person who, in his estimation, is so *mean* as to pursue similar measures in keeping his stock economically.

35. **WINTER FEEDING.**—The Scotch plan of winter feeding is considered a very good one upon mixed arable and dairy farms, which commences about the middle of October, and is often after the following method, the cows being tied up in pairs in the stalls :—

At 8 a.m. each cow gets boiled food, consisting of 30 lbs. of swedes, 1¼ lbs. of linseed, 2 lbs. of bean or pea meal, and a quantity of chaff and light grain unfit for making meal, a liberal supply of oat straw being given after this is finished. At 10 a.m. 60 lbs. of yellow turnips, and oat straw as before. At 2 p.m. about the sixth of a bushel of brewer's or distiller's grains, and at 5 p.m. 60 lbs. of yellow turnips, and oat straw as before; which is the last time they are fed. The accompanying illustration shows a turnip cutter and slicer, manufactured by Messrs. R. Hornsby and Sons, Grantham.

36. **STEAMED FOOD FOR COWS.**—It has been found very advantageous in winter feeding to steam food for cows, some particulars of which we furnish. Mr. Horsfall's management of steamed food for cows has been described in the Journal of the Royal Agricultural Society as follows :—

"The cows are given rape-cake of the kind termed "green" cake, which imparts to the butter a finer flavour than any other kind of cake; and in order to induce them to eat it, Mr. Horsfall blended it with one quarter the quantity of malt dust, one quarter of bran, and twice the quantity of a mixture in equal proportions of bean-straw, oat-straw, and oat shells, all well mixed up together, moistened, and steamed for one hour. This steamed food had a very fragrant odour, and was much relished by the cattle; it was given warm three times a day, at the rate of about 7 lbs. to each cow, or 21 lbs. daily. Bean-meal was also scattered dry over the steamed food, cows in full milk getting 2 lbs. per day, the others but little. He found this substance to be an unfailing means of keeping up the condition of cows while giving milk. When the animals had eaten up this steamed food and bean-meal, they were each supplied daily with 28 to 35 lbs. of cabbage, from October to December (if kohl-rabi, till February) or of mangolds till grass time; each cow having given to her, after each of the three feedings, 4 lbs. of meadow hay, or 12 lbs. daily. The roots were not cut, but given whole. The animals were twice a day allowed to drink as much water as they desired. After the date of his original report, Mr. Horsfall discontinued the use of bean-meal owing to its comparative dearness of price, and gave, in its place, along with about 5 lbs. of rape-cake, an additional allowance of malt-combes, and 2 or 3 lbs. of Indian corn meal per cow. On this food, in instances actually observed, his cows gave 14 quarts of milk a day, at the same time that they gained flesh at the rate of about a ¼ cwt. per month.

A correspondent of the *Agricultural Gazette* upon one occasion described the method he pursued of giving steamed food to his cows :—

" I have a boiler containing about 40 gallons, and into it I put about 50 lbs. of turnips, a considerable quantity of water, and about 12 lbs. of straw cut into chaff, and this is boiled for about two hours, when it becomes a dark, nasty-looking mess; one half of this is taken out into two tubs, and whilst warm 1½ lbs. of bean or pea meal is stirred into each, and then given to each cow at about 110° of heat. That which is left in the boiler remains till morning, and if well covered up is still warm enough for use; it is then mixed with the pea or bean-meal, as before, and given to the cows at break of day; this, with hay *quantum suff.*, constitutes their daily diet, and I get about 6¼ lbs. of butter weekly from each cow. The butter produced in this way has no taste of turnips; and the avidity with which the cows eat this boiled mess is a good criterion of its value. When given to the cows it should be weak and sloppy."

The quantity of butter yielded per cow is not, as will be seen, very large in this instance; but then the amount of feeding fell very short of what is given to cows in order to stimulate a large yield of milk. A useful idea is, however, to be derived from the plan followed of steaming.

Mr. Dancock, of Brompton, a successful manager of cows, whose practice is quoted by Mr. Morton in the Journal of the Royal Agricultural Society, uses steam-prepared food for his cows, giving his meal in the form of gruel over cut hay, or grains, as follows:—

" My plan," says Mr. Dancock, "is to fill with cold water an 8-gallon churn (holding twice that number of imperial gallons) up to the figure 7. This allows room for meal and steam. I then put the steam-pipe within six inches of the bottom, and, supposing the pressure in the boiler to be 10 lbs. turn on full, and in five or six minutes the can is full and the gruel is done. I have sixteen cows, and my quantity is three cans, which allows one large pailful to each cow twice a day. I think this better than giving them meal dry over grains. I milk before feeding, give one bushel of grains to a pair of cows twice daily with gruel over it, and when this is done, give them green stuff and mangolds, a little hay if necessary, then water, and rest till milking-time again, when they are fed as before with grains; then I give oil-cake, about 3 lbs., between two cows, then water, and do up with hay. In the management of cows cleanliness is assential to health—whitewashed walls, mangers well cleaned, cows well cleaned and littered down with straw—in fact, everything belonging to cows and a dairy must be thoroughly clean to preserve health. This, combined with energy and attention. will, in due time, bring profit to the owner."

37. **METHODS OF FEEDING FOLLOWED BY SOME LONDON COWKEEPERS.**—One of the large London cow-shed proprietors, who usually milked thirty cows, has described the routine of the daily work followed.

"At 4 a.m. the cowmen enter the shed, and proceed to milk. In the case of the wholesale milk trade, when the dealers who buy the milk do the milking, one good man suffices for thirty cows. The cowman then only helps, if necessary, at milking-time, and sees that the work is thoroughly done. If he has any reason to suspect that a cow has not been thoroughly milked out, it is his duty to his master to "strip her;" for nothing, as we have previously pointed out, injures a cow more than imperfect milking; and if he succeeds in getting another half-pint from her, his master will give him 6d. for it, and fine the dealer that amount for his servant's default. The milking is begun at 4 a.m., and finished between 5 and 6 a.m. About a bushel and a half of grains is then given between each pair of cows, and they are partly cleaned out, and when the grains are done, a truss of hay (56 lbs.) is divided amongst 12. After breakfast-time, a bushel of chopped mangolds, weighing 50 or 60 lbs., is given to each 2 cows, and the cows receive another truss of hay amongst 12. The cow-shed is then cleaned out, and the cows are bedded, and left. At 1 p.m. milking recommences, and very much the same feeding as previously is given. At 2.30 grains are given as before, followed by the same quantity of hay, and then, being the only time during the 24 hours, the cows are freely watered. They again receive a truss of hay between 12, and are left for the night. The grains are either brewers' **or**

POLLED NORFOLK BREED (FATTENED).

SWISS BREED.

distillers' grains. The former are much inferior to the latter in value, and are less in price ; brewers' grains costing 3*d.* to 4*d.* per bushel, while distillers' are 8*d.* or 9*d.* In the case of cows in heavy milk, and also, for opposite reasons, in the case of those rapidly losing their milk, which must be sent to market as quickly as possible, it is common to give two or three quarts of pea-meal mixed up with the grains morning and evening ; each cow thus receiving that quantity daily. When the milking is coming to an end, for three or four weeks before the cow is sold she may receive 2 or 3 lbs. of oil-cake in addition. A full bushel of grains, half a bushel of mangolds, one-third of a truss of hay, and 5 or 6 lbs. of pea-meal in the case of the fatting cow, are thus the daily ration in a London cow-house. The grains at 2*s.* per quarter, the hay at £5 per ton, and the mangolds at 20*s.* per ton, cost 1*s.* 3*d.* per day, and with meal or cake the daily allowance may cost from 1*s.* 6*d.* to 1*s.* 9*d.* per cow—*i.e.*, 10*s.* to 12*s.* per week.

"The difference in the cost of food in London must be taken into account ; as, for example, mangolds would not be reckoned worth more than 10s. per ton in the country, carriage forming a large item in the cost of London food. Grains in the country are often sold at 6*d.* per bushel (ordinary brewers' grains), which would amount to 4*s.* per quarter instead of 2*s.* In country places these are generally bought and consumed by cottagers for the use of their pigs, and are, perhaps, seldom used to any large extent for feeding cows.

"In summer time in London the cow's food is grass with grains, and meal if necessary. Most cowkeepers, except the very smallest, either have a small suburban farm, or buy a few acres of vetches, clover, or grass, and cart it for themselves. When it is bought daily at the cow-house it costs from 1*s.* to 1*s.* 3*d.* per cwt. during the summer, and the cows receive about that quantity daily, given to them as fast as they can eat it, morning and evening, with their grains.

"Some cows when first put upon grains are very greedy for them, especially distillers' grains, and they yield a large supply of milk upon them, but they soon get surfeited, and it is a bad thing to allow them to have too much at once, it not being wise to allow any description of food to pall upon a cow's taste."

38. DIFFERENT EXAMPLES OF FEEDING AND MANAGE-MENT.

—Mr. John Chalmers Morton points out, in a paper on " Town Milk," contributed to the Journal of the Royal Agricultural Society, several of the facts to which we refer, and remarks that the suburban cowkeeper, though more favourably situated than the London dairy-man as regards the bulk of the food he consumes, the grass, the mangolds, and the hay, is less favourably situated as regards grains ; and this disadvantage, combined with the other of distance from the consumer, is such as at least to balance, often to over-balance, any advantage he possesses over the town dairyman in respect of labour, rent, and cheaper farm produce.

Going further a-field, as, for example, to Swindon and beyond it, or to distant stations on the South Western and North Western Railways, you find that the farmer feeds his cows for London just as he has hitherto done for cheese or butter dairying. Bringing them to the pail at all months of the year, so as to have a regular produce to meet his contract with the London dealer, he milks his cows out at pasture during the summer, and feeds them on hay and mangolds in the winter. Receiving 6½*d.* to 8*d.* per imperial gallon for the milk delivered at the nearest station, and getting 500 to 550 gallons from his cow per annum, he receives £15 to £18 per annum for her produce, which is more than he can generally make of it in the form of cheese or butter, at the same time that he avoids all the cost of labour in the dairy.

AUVERGNAT BREED.

CHAROLAISE BREED.

Milk being sent up to town in this way runs the risk of souring on its journey, in which case it is thrown away at the sender's expense. By cooling it before it starts, this risk is very much diminished; and this is done either by standing the full can in running water, or by placing the milk, before filling it into these cans, in large tin vessels, surrounded by cold water, and traversed by cold water pipes. The risk is further diminished by filling the cans or "churns," as they are technically called, so that they do not shake when travelling on their journey, covering them with wetted jackets, so that evaporation may help to keep the contents cool.

Another Method of Management is thus described by Mr. Collinson Hall, of Navestock, near Brentwood :—

" We begin milking at 1 o'clock in the morning; each man should have 15 cows. The milk arrives at 4 o'clock in London. The cows are again milked at 10 o'clock, and the milk is in London at 1 o'clock. They are fed as follows: Each man gives about 4 lbs. of meadow hay to his 15 cows after the midnight milking, and then goes to bed. At 7 o'clock he gives them half a bushel of grains mixed with a bushel of sweet chaff and a handful of salt; the cows are then cleaned and fresh littered; 2 lbs. of hay a-piece are given, and at 11 o'clock one bushel of mangolds are given; at 4 o'clock p.m. 1 bushel of grains and chaff, and at 6, about 2 lbs. or 3 lbs. of hay.

" The cows are not untied, that they may not mix together, and their water is carried to them. We feed often, and avoid giving large quantities at once.

" Lime on the floors, gas-tar enough not to be offensive, and 10 drops of arsenicum (3rd dilution) in the drinking-water, great cleanliness, and all the provender good, not putting too many in one shed, good ventilation at the top, no draughts : these are my precautions."

The manager of Lord Granville's dairy farm at Golder's Green, on the Finchley Road, in evidence before the Royal Commissioners on the Cattle Plague, described the management of his cows thus :—

" We give about a bushel and a quarter, or from that to a bushel and a half of brewers' grains to each cow, and about 15 lbs. of hay, and about 30 lbs. of mangold-wurzels, with 4 lbs. of meal (pea-meal principally) in addition to that feed in winter. In the summer, grass is given, instead of hay and mangolds. This mode of feeding, though it damages the constitution of a cow, is adopted in order to force the greatest quantity of milk which the dairyman can get. The gain more than covers all the loss; at least, it is supposed to do so. In our suburban district we give them more air, and feed them more on grass in the fields. We do not feed them so heavily upon grains and artificial food as they do in London. We give them much more natural food. Some turn them out from about July to October, and some do not. The cows always lose condition by being turned out ; that is invariably the case. They lose milk, too, to the extent of a quart a-day, unless the pasture is very good indeed."

The allowance of grains we should consider extremely liberal, that is here spoken of as a moderate feed.

Mr. Balls, who managed the dairy farm at Oakington, near Sudbury, in the occupation of Colonel the Hon. W. P. Talbot, has kept from 80 to 100 cows constantly in stalls. They are milked at 3 and 4 a.m., and again at 1 and 2 p.m., and are fed exactly on the London plan, first on grains, a bushel between two, next with a little hay, then with a bushel of either cabbages or mangolds, and then again a little hay; in the afternoon, grains, and hay, and water (they are only watered once a-day), and again hay before night. The alteration in summer is a substitution of grass for hay and mangolds. A small quantity (3 or 4 lbs. a-day) of meal is given, along with grains in the case of cows nearly dry; or rather this used to be given, for Mr. Balls declared that there was no profit in the attempt to put on extra flesh with extra feeding, while meal was dear and meat so

TURNIP PULPER.

cheap. Meat, however, while this is being written, is very high, while it was very low at the time Mr. Balls was speaking. He contrived, however, by careful purchasing, to get cows which would put on flesh without extra feeding when they got dry.

The Turnip Pulper shown in our illustration is that supplied by Messrs. R. Hornsby and Sons, Grantham.

39. SHORTENING THE COW'S SUPPLY OF FOOD BEFORE CALVING.—It is a very common error with many cowkeepers to shorten the supply of food to the cow during the time she is dry before calving. This is a great mistake, as it tends to weaken the

cow when she has most need for all her strength, and it frequently happens that this course has the effect of lessening the supply of milk after she has calved, till she becomes dry again, while it doubtless injures the calf. On the other hand, if the mother is well fed up to her time of calving, her progeny will be strong and healthy.

40. FEEDING COWS FOR MILK OR BUTTER.—What cows are fed upon makes a considerable difference in the results, and the appropriateness of the method adopted in feeding. If the production of butter is intended, or whether milk alone is sought to be produced for sale, it makes all the difference as to the kind of food which is given to them.

During the winter and spring months, when roots form a great proportion of the food which is given to cows, some of them are apt to impart a disagreeable taste to butter, injuring its sale. The most commonly objected to, and that most widely known, is the taste of turnips, which is particularly offensive and disagreeable to some people. Yet the butter made from the milk of cows fed on turnips can be had perfectly sweet and good if certain precautions are used. A common practice exists to obviate this by putting saltpetre into the pans; but the unpleasant flavour arising from turnip-feeding may be counteracted by giving the cows a small quantity of concentrated food, the most suitable of which are crushed oats, beans, Indian and palm-nut meals, bran, and oil-cake.

41. A COURSE OF GOOD FEEDING HIGHLY REMUNERATIVE.—It is a matter of experience with those who keep milking-cows, that the better the animals are fed, the more remunerative they become; and it pays well to give them linseed, or rape-cake sometimes, in addition to the best food which can be obtained from the farm. Two pounds of rape-cake will cost about twopence for each cow daily, and an increase of one pound of butter per week may be reckoned upon, besides keeping the cow in vigorous health, which a little generous diet will tend greatly to effect, as it will have a beneficial effect upon the rest of her food. The advantage of giving something of this sort constantly, also, will neutralise the ill effects which a change of food dependent upon the seasons will sometimes bring on. As the balance between loss and profit lies in giving just sufficient for the purpose, and no more, care should be taken that these artificial aids should not be administered extravagantly, or of course they will become too costly. Many farmers, who are quite alive to the good effect resulting from this course of treatment, have

discontinued it on account of the extra expense incurred, the food having been given wastefully; but, used with proper caution as auxiliaries, they will be found to well repay the outlay.

42. **WATER.**—It is scarcely necessary to point out that cows should have a regular and sufficient supply of clean water. Many cows will of themselves seem to prefer even, and drink, the fetid water that sometimes accumulates on the surface, into which the drainings from a manure heap have flowed; and they should, there-fore, never be allowed to have access to foul water, if there is any means of preventing them.

MILKING PAIL.

CHAPTER IV.

DISEASES OF COWS.

Catarrh—Diseases of the Organs of Respiration—Bronchitis in Cattle—Hoove, Hooven, or Blasting—Choking—Distension of the Rumen with Food—Loss of Cud—Inflammation of the Rumen—Moor-ill and Wood-evil—Scouring—The Scant—Diarrhœa—Redwater—Retention or Stoppage of the Urine—Diseases of the Udder—Rheumatism—Cow-pox—The Drop—Abortion—Slinking—Slipping Calf—Warping—Inversion of the Uterus—Shelter for Cows—General Hints upon the Management of Cows—A Clergyman's Experiment.

43. DISEASE is very much influenced by climate and the season of the year, the result being that, in warm weather, affections of the digestive and abdominal organs are the most frequent; whilst in cold weather affections of the chest, rheumatism, and kindred ailments which arise from it, are sometimes common, especially when animals are not provided with sufficient shelter, inclemency of the weather inducing epizootic and endemic diseases.

44. CATARRH: DISEASES OF THE ORGANS OF RESPIRA-TION.—These prevail mostly in the spring of the year, when the wind is easterly, and particularly if the weather is both cold and wet. Stock also are subject to attacks in wet weather in the autumn, the young animals being more sensitive to this, as well as to other diseases affecting the air passages, than older beasts.

Some warm bran mashes, with a little nitre in them, is good treatment in mild cases, and will generally be found efficacious; but in a severe case, bleeding, and a dose of Epsom salts, are prescribed; a stimulating liniment rubbed into the throat, or a seton may be inserted.

The following is a good liniment to rub into the coarse skins of cattle when an external stimulant is necessary:—Powdered cantharides, 1 oz.; olive oil, 6 oz.; oil of turpentine, 2 oz. Mixed together.

When catarrh assumes an epidemic form it is desirable to use vegetable tonics, such as ginger and gentian-root, as there is greater tendency to debility, and it is generally more severe.

45. BRONCHITIS.—Neglected catarrh will often bring on bronchitis in cattle, which results from extended inflammation over the same membrane to a more dangerous part on the internal surface of the lungs. The symptoms are similar to those of severe catarrh, but the animal experiences greater soreness in the act of coughing. Bleeding should be resorted to in the early stage of the disease; a seton should be inserted in the brisket, and mild aperient febrifuge medicine administered.

46. HOOVE, HOOVEN, OR BLASTING.—Meteorization, which passes generally under one or other of the above names, is literally distension of the rumen with gas given off by the food taken by the animal, which has fermented, and the stomach is soon distended to an enormous size. Cattle which have sometimes broken loose, and have trespassed on a clover field or other green crop, and have eaten inordinately, are very liable to it, and suffocation will take place (from the carburetted hydrogen in the early stage, and afterwards the sulphuretted hydrogen), if relief is not soon afforded.

The treatment is to liberate the confined gases, or to condense them by chemical re-agents; and to do this the hollow flexible probang should be passed down into the stomach, so that the gas may escape through it.

Either before or after this operation the following draught may be given:—Powdered ginger, 3 dr.; hartshorn, 1 oz.; water, 1 pint.

If these ingredients should not be at hand, two drachms of chloride of lime, dissolved in a quart of water, should be given, or some lime-water, which, however, is not so efficacious. A purgative should be given after these medicines to restore the power of the digestive organs.

At an advanced stage, it is sometimes necessary, in order to save life, to make an incision in the flank, on the left side, between the last rib and the hip-bone. An instrument termed a trochar, which is inserted in a tube called a canula, is employed for doing this, the former being withdrawn, and the latter retained until all the gas has escaped; but if this is not ready at hand, a pen-knife may be used, and a quill, or stick of elder can be substituted; the small wound being afterwards closed with a stitch and a bit of plaister.

47. CHOKING.—A good many animals are lost from this cause in the course of the year; a piece of turnip, a potato, or a stray apple

which has been picked up, and hastily swallowed, becomes impacted in the œsophagus, and pressing in the softest part of the wind-pipe, interrupts respiration; and if not removed in time, ends in suffocation. Sometimes, in the removal of the obstructing object, the œsophagus is so lacerated that the animal never recovers, a smooth object being more dangerous than an irregular one.

The best treatment for this injury is to administer a little oil or lard, by the horn ; a rather large probang, with a knob at the end cut obliquely, should be passed along the roof of the mouth till it enters the œsophagus. When the obstructing body is touched, the head should be alternately raised and depressed, and only moderate pressure of the probang resorted to. If it does not readily pass, it is better to wait a little rather than use force and violence, which has been the occasion frequently of killing animals, and try again a little while after. The longer the obstructing object remains, the softer it gets, and a second time it may be removed very easily. Too great force, when used, will lacerate the lining membrane of the œsophagus and its muscles, as will ragged tube-ends. Laceration is evidenced, when in the neck part, by a swelling which hourly increases, generally above the occident, in much greater proportion than below. The swelling is hard and tense, from an infiltration of mucus into the surrounding tissues. Fever sets in, and respiration becomes painful. The animal moans, and refuses everything. The breath becomes fetid, and death often ensues from the third to fifth day. As the poor beasts generally die after this laceration, if they are in good enough condition for the butcher, it is thought better to slaughter them at once, and not wait for the further development of the injury. When the animal makes an attempt at vomiting, it usually denotes an obstruction near the entrance of the rumen, when the obstructing body can only be removed by drawing it upwards, this being particularly the case when it is impacted in the roof of the mouth, which will be shown by an uneasy motion of the head and a discharge of saliva from the mouth. The object in this case is best removed by the hand, though sometimes considerable force is required. When these means fail, rather than use too much violence, when meteorization, or hooven, is produced by choking, it is preferable to open the œsophagus and remove the obstructing body. The operation is termed œsophagotomy, and is best performed by a veterinary surgeon, rather than an unskilful person.

48. DISTENSION OF THE RUMEN WITH FOOD.—This, though not attended with such acute symptoms in the early stage as

D

hoove, is more difficult to remove, but is fortunately of rarer occurrence, happening mostly with stall beasts; but it is important to
distinguish between distension with gas, and with food, although it
is somewhat difficult to do so, the symptoms being similar.

The distension produced by solid matter is not so great, nor the
distress so urgent, though the danger may be sometimes greater.
The abdomen feels hard in the region of the rumen, and if the probang is used, there is no gas liberated.

In tympanitis, from overloaded stomach, meteorization is often
the first symptom, as well as fulness and hardness of the paunch;
often the seat and source of the inflammation of the powers of
digestion. This variety resists the power of mucilaginous drinks, of
ammonia, and other remedies, and even of puncture.

The hard and dried accumulated food in the rumen forms certain
pellets, which, on account of their bulk, can no longer be returned
to the mouth for a second mastication. The contents of the rumen
should be ascertained by means of the trochar; and also to what
extent the distension exists, which can be discovered by the resist-
ance offered to the trochar in moving it to and fro.

49. LOSS OF CUD.—This is more frequently a symptom of
disease than a disease itself, though it is a proof that there is considerable derangement of the bodily functions; and the resumption
of rumination is justly regarded, in cases of illness, as a sign of
approaching convalescence. When loss of cud occurs without any
traces of decided disease, it is best treated by mild purgatives and
stomachics.

50. INFLAMMATION OF THE RUMEN. — When poisonous
plants prevail extensively in a pasture, such as hemlock, water-
dropwort, henbane, wild parsley, or even the wild poppy and the
common crowfoot, inflammation of the rumen will sometimes be
produced, but the cases are extremely rare, as the fine sense of
smell with which cattle are endowed enables them to reject those
plants that are inimical to health, though they will eat the yew
(which is most fatal when withered and dried) from the clippings of
trees which have fallen into their pasture.

The effects of this poisoning are usually of a narcotic character,
and a change of pasture should be made, and medicine of a purgative character administered.

These narcotic plants, taken with the food, will affect the second
stomach, or reticulum, of the animal; but much more frequently the
maniplus, or manifolds; and under the term of "Fardelbound" is

an ailment arising from a retention of food in this stomach between its numerous plaits. Too much food of a dry character, and insufficient moisture, tend to this, as well as other causes, to derange the digestive organs.

But the same appearance of the maniplus is also found connected with other diseases, and this constipated state is occasionally found when the bowels are quite relaxed. Aperients, combined with stomachics, is the best treatment to resort to—Epsom salts with ginger, in applicable doses, being the most convenient form.

51. MOOR-ILL, AND WOOD-EVIL.—In dry seasons a disease is met with, most frequently in the neighbourhood of woods and commons, when the secretion of milk is lessened, and the animal refuses to eat her usual quantity of food. The appetite is at best capricious, and the cow will pick up stones, pieces of bone, or iron, and will grind them in her mouth for several hours successively. She will also seize and devour all the linen she can possibly get at, and many a poor washerwoman, drying her clothes on a common, has been scandalized by this erratic behaviour of, perhaps, an ordinarily well-conducted cow. She drinks, also, the oldest and filthiest urine she can find, which she prefers to the purest water.

The earlier symptoms are a staring of the coat, and a seeming adherence of the whole integument of the ribs below, so that it can scarcely be raised by the fingers. The belly is tucked up, and the animal gradually loses flesh, the bowels being confined, from the earliest appearance of the disease to its termination. Constipation is a regular attendant of wood-evil, sometimes very obstinately so. Stiffness supervenes in various parts of the body, but more commonly in the fore extremities, the shoulders, or the chest; often shifting from limb to limb. Sometimes great lameness will ensue, and this, in the same way, shifting from joint to joint. When the animal is induced to move, she utters a kind of interrupted moan, or groan, expressive of the pain she is experiencing. There is also a singular cracking noise to be heard when she walks, as if the bones of the joints were moving in and out of the sockets at every step she took. The animal begins to heave at the flanks, sometimes very violently, and the pulse is accelerated at times to more than 100 beats a-minute; the bowels, which all along have been confined, get more so as the disease proceeds. The secretion of milk almost ceases. The animal seldom ruminates, and can be scarcely induced to eat anything.

The proper treatment in the first place is to give a good strong dose of aloes in solution, and regulate the bowels, which, if it does not produce the desired effect, must be followed up by salts, repeated every six hours till they operate. Bleeding should not be resorted to, unless there are symptoms of inflammation of the lungs, in which case it is desirable, and will relieve the animal very much; but this must be practised with caution. The aperients should be followed up with febrifuge and alterative medicine, until the organs of digestion are restored to their natural tone, the diet consisting of mashes and gruel. In addition to this plan of treatment, a seton is sometimes inserted in the dewlap, and, in very severe cases, as much as 10 lbs. of blood have been taken away; and 6 drachms of aloes, 12 ounces of sulphur, 16 drachms of croton oil, with 3 drachms of powdered carraway seeds, administered; the second day 8 lbs. of blood removed, with repeated purgatives in lessened quantities, blistering the animal's sides as well.

52. SCOURING; THE SCANT; DIARRHŒA.—The symptoms which denote this disease may proceed from various causes, the relaxed state of the mucous coat of the small intestines being amongst the most simple. In severe cases this may proceed from disease of the liver, stomach, or maniplus; and when the diarrhœa is produced by unwholesome food, a change of diet will sometimes effect a cure, but if it does not cease, the following is a good astringent and tonic:—Prepared chalk, 2 oz.; gentian root, powdered, 2 dr.; opium, powdered, ½ dr. This should be well mixed with thick gruel, and given once or twice a-day, as required. If the animal is very young, a smaller dose should be given.

Should, however, the liver be affected, calomel in combination with opium is more to be relied on; half-a-drachm of each being given twice a-day. In bad cases it is good practice to clear out the intestines by a dose of salts, and afterwards give the calomel and opium.

53. REDWATER.—Redwater is a disease of the digestive organs, and principally of the liver, the urine being loaded with biliary deposits, which should have passed away by other channels.

Formerly it was regarded as disease of the kidneys, the dark colour of the urine being attributed to the presence of blood. It is frequent in cows several weeks after parturition. The first symptoms are diarrhœa, soon succeeded by constipation. The appetite falls off, and the pulse and breathing get accelerated, the former weak, with coldness of the extremities. Rumination ceases, and the milk is diminished, the urine becoming brown, and sometimes even black.

The disease is most prevalent after hot, or dry weather; and is sometimes brought about by the change from a poor to a rich pasture; and from marshy and cold to dry, sweet, and stimulating pastures, especially in elevated situations. It is commonly supposed that to take a cow from an inferior pasture, and put her into a good one, is the way to improve her health and increase her produce, and so it will ultimately, but like sudden changes in the human animal, from temperate or spare diet to unaccustomed rich eating and drinking, at first the system is likely to be deranged by it.

The remedy consists in opening the bowels, for which the following is well adapted:—Sulphate of magnesia, 12 oz.; sulphur, 4 oz.; carbonate of ammonia, 4 dr.; powdered ginger, 3 dr.; calomel, 1 scruple; made up into a draught, with warm gruel. One-fourth of the above may be given every six hours; after which, mild stimulants, with diuretics, may be given, as the annexed:—Spirit of nitrous ether, 1 oz.; sulphate of potash, 2 dr.; ginger, 1 dr.; gentian root, 1 dr. To be given twice a day.

54. **RETENTION OR STOPPAGE OF THE URINE.**—This sometimes occurs with pregnant cows, and arises from a pressure of the womb on the stomach. The urine needs to be removed by means of a hollow tube, called a catheter, and the other symptoms which may attend this derangement should be treated according to their several exigencies.

55. **DISEASES OF THE UDDER.**—The udder of the cow is subject to attacks of inflammation, particularly after calving, when it swells, feels hot, and the part affected becomes hard. The secretion of milk is also interrupted. In this condition it is termed *gargel*. Sometimes exposure to cold and wet will bring it on, and in severe cases the cow will lose one, or two quarters of the udder, and occasionally these cases end fatally.

Hot fomentations should be applied in the first place, and if the inflammation is excessive, bleeding from the milk-veins of the affected side should be adopted. A purgative also will be found useful, but if the complaint commences with shivering, a stimulant is necessary, such as an ounce of ginger dissolved in warm gruel, or ale, with two ounces of spirits of nitrous ether, which will, sometimes, at once stop the progress of the disease.

After fomentation, an ointment composed of the following may be rubbed into the udder, on the part affected:—Camphor powdered, 1 oz.; mercurial ointment, 2 dr.; lard, 8 oz., well incorporated together.

56. **RHEUMATISM.**—Joint Felon and Chine Felon are common

terms for rheumatism, generally produced by exposure to the weather and careless treatment, which may be either general, partial, severe, or sub-acute. The fibrous tissues become affected, and it may either affect the muscles or sinews, or extend itself to the serous membrane lining the chest, and investing the heart. Its presence is indicated by great pain, and stiffness in moving, attended with considerable fever; but when the attack is sub-acute, the joints are generally affected. The common treatment is to bleed in the first instance, followed by a purgative, and with it an ounce of the spirit of nitrous ether. This may be given twice a-day, with a drachm of tartarised antimony, and one of colchicum.

The parts principally affected may, with advantage, be fomented, and afterwards well rubbed with a stimulating liniment.

57. COW-POX.—The Cow-pox is not by any means a common disease, and consists of the formation of numerous pustules on the udder and teats, the contents of which are infectious, as is well known in the case of the human subject, where vaccine lymph is employed, and it may be propagated by the hands of the milker, from one cow to another.

A cooling aperient should be given, and a weak astringent applied to the sores on the teats. This can be made with a little powdered chalk, with one-fourth part of alum, which will be found a very useful application, the treatment being simple enough.

58. THE DROP.—This disease seldom takes place until the cow has had several calves, and is supposed to arise from a depression of the nervous system, caused by the after-pains, or reaction of the womb after birth, which, added to the previous muscular efforts in expelling the fœtus, produce exhaustion; the nerves devoted to these organs, and the spinal marrow at the region of the loins, becoming over-taxed.

With each successive calf, the uterus becomes more dilated, and, consequently, the contractions afterwards are greater, and more attended with danger, than when the cow has her first calves. It is therefore often very annoying to find a fine cow, which has brought a good calf, and is apparently doing well, attacked by this disease finally, which literally lays her low. There are two varieties of the disease, one acute, the other sub-acute. In the one it is generally fatal, the other being usually curable, the former being characterised by utter prostration of the vital powers, while in the other some degree of animation and appetite is retained, though without the power to rise, or stand.

One of the symptoms is a torpid state of the bowels and stomach; rumination ceases, and the food in the various stomachs remains in an unchanged state. Purgative stimulants should, therefore, be applied. The cow in an acute variety of this disease can take a large amount of medicine, as much as the following:—Sulphate of magnesia, 1 lb.; flowers of sulphur, 4 oz.; croton oil, 10 drops; carbonate of ammonia, 4 dr.; powdered ginger, 4 dr.; spirit of nitrous ether, 1 oz. The above should be dissolved in warm oatmeal gruel, and given slowly and carefully to the animal. In unusually severe cases, the croton oil can be increased; and from four to eight grains of powdered cantharides may be added. A strong blistering stimulant should be rubbed over the spine and loins, and a fresh sheepskin, with the wool outwards, has, with advantage, been placed on the loins of a cow so affected. Every six hours, one-fourth of the above medicine should be given, with the exception of the croton oil, until purging is produced, and if the cow cannot pass her urine, it should be removed by means of the catheter.

In the milder forms of the disease the medicine should be administered in greater moderation; but as prevention is better than cure, in-calf cows should have plenty of exercise, shelter from the weather, and moderate feeding, but not too low feeding, which we have spoken of before. If, however, there is reason to expect a cow may be subject to the disease, it is better not to feed too heavily.

Confinement to the stalls is a bad practice *before* calving, though it may be done with impunity *after*. Sufficient nourishment is necessary for the cow, but the stomach must not be overloaded so as to press upon the womb; and for the proper motion and health of the fœtus, exercise is strictly necessary. The animal must not make too much flesh.

Particular care should be paid to the state of the bowels, which should be kept open, and as the period of calving approaches, unless the fæces are much relaxed, one-half of the purgative above described should be administered, and a few bran mashes, instead of the usual quantity of hay, be given, in order to prevent the stomach being overloaded with food difficult of digestion.

If the cow does not clean properly after calving, it is advisable not to be in haste to remove the after-birth by manual operation, but to give the mild purgative before advised, and wait a few days; after which, if it does not come away, the hand should be passed up, and the after-birth removed with as little force as possible.

Care should be taken that the in-calf cow is not worried by dogs, or allowed to leap her fences; and, at the same time, protection from the weather must be afforded at ungenial seasons, without too much confinement.

59. ABORTION; SLINKING; SLIPPING CALF; WARPING.— Abortion in the cow commonly takes place between the ninth and fifteenth week, but it may occur at any period of pregnancy, the cow being supposed to go with young about nine calendar months, or 284 days, though the period is more often exceeded than the contrary.

Its occurrence is conspicuous at particular seasons, as if there was some unseen connection with the atmosphere, being more frequent after the prevalence of wet weather. The ergot of rye has a very exciting effect upon the uterus, and as rye grass, and grain, are subject to the same disease, it has been considered, with much plausibility, that the unusual presence of this poisonous matter in the grasses has, at times, a great deal to do with abortion. It is said also that the smell of a cow which has aborted has a tendency to produce the same effect upon another pregnant animal.

A cow that has warped once is liable to do so again, and there is danger of the mischief spreading; it having, at times, been necessary to get rid of a large herd from this cause. Cows that do not breed early are more likely to abort than those which are put to the bull as soon as the inclination shows itself.

At an early stage of pregnancy, when abortion takes place, there is little disturbance to health, and treatment is seldom required; but at a late period, serious consequences, such as inflammation of the womb, and even death, follow.

Abortion may be brought about by blows, strains, or even jumping, or riding other cows—from fright, or excitement of any kind, as well as by disturbance of the digestive organs. Some times the causes are of a constitutional nature, and arise from some hidden defect in the procreative organs, high-bred animals in high condition being more liable to this than others.

When treatment is required, a dose of salts should be given to relax the bowels, which may be followed by a sedative, such as an ounce each of laudanum and spirits of nitrous ether. Where there is inflammation of the womb, hot fomentations should be applied externally to the loins, for a good stretch of time together, and warm water is sometimes prescribed, to be syringed into the blood. Bleeding is also occasionally resorted to.

Prompt treatment will often stave off threatened abortion. The cow should be kept quiet, and bled, and one and a half ounces of tincture of opium, and the same quantity of spirit of nitrous ether given; but no purgatives administered. If a cow has aborted before at a particular period, it is a good precaution, and is considered prudent, to bleed her just before this time.

It generally happens that the after-birth is retained after abortion; and it is the best course to remove it, although it may be necessary to introduce the hand into the uterus, and take away the placenta from it, by carefully breaking down the points of attachment.

60. **INVERSION OF THE UTERUS.**—Both inversion of the uterus, and inversion of the vagina, take place occasionally; the

SHORTHORN BREED.

GALLOWAY BREED.

former being the most serious, and generally occurring after parturition. In both emergencies, the parts should be carefully cleansed, and returned as quickly as possible, and a bandage applied, the hind parts being kept higher than the fore ones.

Instances have been known of the inversion of the vagina, produced by violence, which have been successfully reduced, and a healthy calf dropped a few days afterwards.

Unnatural presentation will sometimes prevent a cow from calving, or a scirrhous state of the mouth of the uterus. The proper treatment in the last instance is to divide the stricture carefully. In the case of unnatural presentation, endeavours should be made to return the calf to its former position, which is with the head resting on the fore-legs, that ought to come first. In some instances it is necessary to turn the calf. When the hind parts come first, care should be taken that both feet emerge before the buttocks. Sometimes in these unnatural presentations considerable force may be necessary to assist labour, and in some very bad cases it may be obligatory to take away the fœtus piecemeal, in order to save the life of the mother.

61. SHELTER FOR COWS.—From the foregoing list of ailments to which cows are subject, it will be seen that exposure to the inclemency of the weather is a fruitful source of disease in one form or another to them; and that occasional shelter is absolutely necessary, though in some grazing counties, as Gloucestershire, very little is provided for them.

It is especially an important matter to keep milch cows warm in winter, and one of the things which strike a stranger very forcibly when he enters a London cow-house during cold weather, is the warm temperature the cows are kept in. Experience has shown that this has an important influence on their productiveness. They stand very thickly on the ground, one to every 30 to 36 square feet, where the closeness with which they stand causes warmth, and the windows are closed and matted, and no thorough draught allowed, and thus the shed is kept warm. There is generally room enough overhead, and perhaps a tiled roof, which allows ample ventilation; and thus, where the shed is kept clean, the air is sweet enough, as well as warm. If not to be exactly imitated, a useful hint is to be gathered from this method of procedure by the country dairy-farmer.

62. GENERAL HINTS UPON THE MANAGEMENT OF COWS.—Upon the well-known principle of prevention being better than cure, it would perhaps be appropriate if we were to close this section of our work with a few general hints upon the management of cows, as a little attention to details, and careful treatment, will often keep off disease, and when dealt with at an early period, incipient disease can be more easily eradicated than when it has assumed a definite form.

The first indication of failing health on the part of the cow is a falling off of the supply of milk. This will often take place before the appetite of the cow fails. By those who pursue an efficient system, and neglect no precaution in the management of their animals, this symptom is at once noticed.

It may be only a temporary ailment, or it may be the forerunner of more serious disease, and in either case it is thought a good plan to give a drench at once. One ounce of nitre in a quart bottle of water, into which four ounces of flour of sulphur have been shaken, will be found efficacious. Some make a point of giving this mixture to all new animals that are purchased, before they are put with the rest of the stock, from which they are isolated for a few days, so as to give an opportunity of judging whether the fresh arrival is free from disease.

The dry and soft food should be regulated according to the condition of the dung. If a cow becomes costive she loses her milk, so that her dung ought to be rather loose than otherwise, to be an index of her good condition.

Good food and water, regularly given, are the most essential points in feeding cows. It has been proved that cows which have been fed regularly upon inferior food, have yielded more milk than those to which richer food has been given, but not at regular intervals. Irregularity in the hours of feeding is invariably followed by a smaller supply of milk, and where this falling off has taken place it takes some time for the cow to resume giving her proper quantity which she has been accustomed to do with regular feeding.

Common salt, given in moderate quantities to cows, increases the quantity and improves the quality of the milk. About four ounces a-day would be considered a proper quantity; and cows ought always to have ready access to water.

Those cows which are nearly due to calve should be kept separate from the others, which sometimes ride them, when there is a risk of the calf turning in the cow, in which case a bad calving may happen.

It is usual with many to desist from milking about eight weeks before the cow calves. But this depends upon circumstances. With some cows the milk will have become very reduced in quantity, but in others a good flow will continue, in which case it will be expedient to milk once a-day perhaps, or once in two days.

When there are a large number of cows, the heifers ought to be kept by themselves.

As cows frequently manifest a degree of pugnacity, and quarrel with each other, they should be kept, as it were, "assorted" when they are tied up in the yard, commencing with the "best woman" at top, then next best to her, down to the meekest in the herd last, the least able to bear the ill-temper of the strongest, by which

NEW LEICESTER.

DURHAM-CHAROLAISE BREED.

arrangement all will be able to eat their food in greater tranquillity.

In buying a cow, the purchaser should choose one with a large soft udder, and the teats not too close together. The teats should also be of fair size. When taken home, she should be kept separate from the others for a short time, for she may have some latent disease, which time may develop.

Any falling off in the supply of milk an animal has been in the habit of giving should be at once noted, for it is an unfailing indication that there is something amiss; and this will take place sometimes, as before stated, before the animal's appetite falls off.

An irritable cow is generally an inferior milker. An animal with a placid, ruminating disposition yields the most milk.

It will be found a good plan to have some vetches, or other green food ready, when the meadows are parched up with the summer heat, and to keep the cows in sheds, or under some kind of cover, to prevent their being tormented with flies; and let them out only during early morning, and evening.

The more pains and care that are taken, the greater will be the return made in produce, and it is really astonishing what may be done by good and regular feeding, and careful treatment in every way.

A Clergyman's Experiment.—A somewhat whimsical course of experience was undergone upon one occasion by an old friend of the writer's, a clergyman, who had but a very small income, but a large family of boys and girls. He had found no difficulty in educating them, for even his girls were familiarly acquainted with the best Greek and Latin authors, in whom he had instructed them himself; but he had considerable difficulty in finding them sufficient bodily food, though they were mentally so exceedingly well-fed; and being very desirous of increasing his income, having perused a treatise upon a certain Yorkshire cow, in which the owner proved to demonstration that it had been made to yield milk and butter which amounted in value to £2 per week, by a certain course of treatment, he resolved to make the experiment himself, and accordingly bought a Yorkshire cow.

He followed the treatment prescribed most accurately, and the results were certainly wonderful; for he had secured a good animal. But then, as he pathetically added, it took up the whole of his time, and that of the boy who did the work about the place, to wait upon this cow. She was to be fed regularly so many times a-day. Water was to be given to her at due intervals; she was to be curry-combed every now and then, and all her requirements were to be attended to with the greatest precision; so that he found the task too much for him, and this cow at last had to mingle with the common herd of cows, after he had resolved to get rid of her, and was ever afterwards undistinguished, beyond being considered an excellent animal, which she undoubtedly was. We often hear of these wonderful results from individual owners of cows, and the secret is, they receive a much larger share of attention than is bestowed upon average animals. Their example is, however, valuable in showing what may be done by that care and attention.

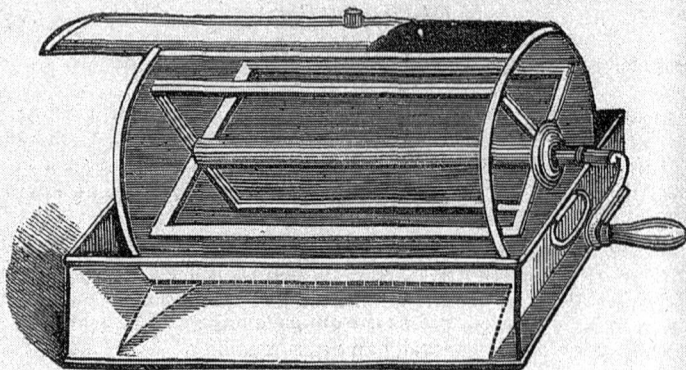

THE "SUSSEX" BUTTER-CHURN.

CHAPTER V.

THE DAIRY.

Situation and Construction of the Dairy—Best Materials for Building—Ventila-
tion and Arrangement—Cleanliness—An Inexpensive Dairy easily Constructed
—Vessels and Implements of the Dairy—Milk-Pans—The Churn—Cheese-
Presses—Shelving of Stone or Slate—Dairymaid's Duties—Dairies in Town.

63. **SITUATION AND CONSTRUCTION OF THE DAIRY.**—The
dairy should be placed tolerably near to the house, for convenience
sake, but should be away from the farm-yard, as well as distant from
any pond, or stagnant water, for milk is soon contaminated by the
near proximity of any decaying matter, and quickly absorbs impurity, and thereby acquires an unpleasant taste.

It is necessary for the dairy to be cool in summer, and warm in
winter; and if the main aspect is open to the north and east, it is
considered best, and shaded from the south and west by trees or
walls.

A Sunken Floor, with a span roof projecting broadly over the side
walls, tends to keep the dairy cool in summer; and thatched roofs
are liked, as they keep out the sun, which often lies hot upon the
tiles with which outbuildings are generally covered in many parts
of the country, and are also warm in winter.

The Thatch should be made of clean, sweet straw, or if the roof
is covered with *thick* slates it would be even better, as sometimes an
old thatch gets unpleasant from decay, and the smell is apt to com-

municate its taint to the milk. The greatest care should be taken to guard against chance of contamination from any source whatever, and there are often unsuspected sources, of which this is one.

64. **BEST MATERIALS FOR BUILDING.** —Slate is the best material that can be used about a dairy, either for shelves, flooring, or sides of the building. Many handsomely-constructed dairies are fitted up with marble, which seems to have become regarded as the best material to use for shelves, but fishmongers find that fish are preserved sweet for twenty-four hours longer on slate than on marble.

A Slate Floor also presents a smooth even surface, from off which any spilled milk can be easily removed. When dairies are paved with brick, spilled milk stands in the interstices, and the sour smell which it creates will impart a taint to that in the pans, notwithstanding the floor may be washed with water, as crannies, or inequalities in the flooring cannot always be reached.

Tiles and Bricks absorb a large quantity of moisture, while slates, it is said, imbibe but the two-hundredth part of their weight, and tiles absorb one-seventh.

65. **VENTILATION AND ARRANGEMENT.**—The dairy should be constructed with sliding windows, or valves, to regulate ventilation and secure a constant supply of fresh air. A churning-house should adjoin, divided from the compartment where the milk stands in flat pans, with a boiler in one corner, fitted with vessels, either of lead or slate, for holding the whey. It will be found a good plan to have a tank or receptacle outside, with a pipe communicating to it, by which the whey can be let off for the use of the pigs. Whey keeps longer sweet in lead than in wooden vessels, but becomes very offensive when any sour liquid is allowed to remain in them; slate, however, is better than either wood or lead.

66. **CLEANLINESS.**—Adjoining the dairy should be a washhouse, containing a pump of good spring water, and also a furnace with cauldron for scalding out all vessels and utensils, so that they may be kept sweet and clean. Plenty of cold water thrown down upon the floor of the dairy in hot weather, keeps it nice and cool. Provision should always be made in the pitch of the floor, or floors if they are separate, so as to allow of all the water draining thoroughly off, carrying away with it all traces of the milk which may have accidentally dropped upon it. A long bench should be placed outside the door of the wash-house, on which the utensils should be put to sweeten, and dry in the sun and air, after being thoroughly

well washed. Badly cleaned vessels are often a source of loss to the owner, and should be carefully guarded against.

67. **AN INEXPENSIVE DAIRY EASILY CONSTRUCTED.—** Where there is not adequate dairy accommodation, an inexpensive dairy can soon be constructed, at a very moderate cost. A frame could easily be put up of light square pieces of wood (called quartering by carpenters or timber-dealers), cased outside with half-inch slates.

ROLLER BUTTER-PRINT.   CREAM-SKIMMER.

PATENT MILK-PAN.

STRONG MILK-PAIL.   SIEVE.

The cavities between the quartering to be filled up with solid concrete, or with rubble of brick and stone, plastered smooth inside with a trowel, and lime-washed. Concrete is easily made, in the proportion of seven measures of gravel to one measure of fine stone quick-lime. A flooring of slate, laid upon this concrete, four inches thick, the slates laid in a bed of mortar, makes one of the best floors it is possible to have, and it can, of course, be sloped in any direction for the purpose of draining off the water which is used to wash it. It must ever be borne in mind that, although water used in plenty for the sake of ensuring cleanliness is highly advantageous, yet water should never be used unnecessarily, as it is highly desirable that the dairy should be as dry as possible, damp being very prejudicial to its operations.

68. **VESSELS AND IMPLEMENTS OF THE DAIRY.**—When dairy operations are conducted upon a large scale, and of a varied nature, there are a good many utensils and implements, of one kind

LAWRENCE'S MILK-REFRIGERATOR.

or another, which need getting together, which may be briefly mentioned as comprising milk-pails, milk-pans, sieves for straining the milk when taken from the cow, cream-pots, or dishes, churns for making butter, scales for weighing, and cut wooden prints, and boards for ornamenting it. When cheese is made, large vessels are required to hold the whey and butter-milk—vats, tubs, curd-breakers, presses, and ladders.

The above sketch represents a milk-cooler, or refrigerator, of which there are various makes and forms.

E

69. **MILK-PANS.**—Much difference of opinion prevails with regard to the kind of pan which is best adapted for containing the milk while held on the shelves of the dairy. In most places they are of wood, though many people make use of earthenware, but wooden

LANCASHIRE PLUNGE-CHURN.

coolers are generally liked the best. They are liable to fall to pieces if kept for a long time without being used, but otherwise they are the most economical, as there is no breakage, and with care they will last a lifetime. When kept perfectly white from assiduous scouring, and the hoops shining like silver—which some dairymaids who

DERBYSHIRE BUTTER-CHURN.

take a pride in their utensils will cause them to look like—they have quite an ornamental appearance in the dairy; while in winter they possess the merit of not cooling the milk too suddenly, which is a qualification highly advantageous to the rising of the cream. There are also iron vessels tinned, as well as of slate and glass. The high price and brittle nature of the last have operated against their extensive use, though they are liked very much. Milk-pans, when of wood, are generally made of the best oak or maple. Shallow pans are supposed by many to be more suitable for setting the milk, and throwing the broadest surface of cream which it is possible to get to the top; but in the height of the season, when the dairy is crowded with standing milk, objections have been made to the extra room taken up by flat dishes; while in winter, it is thought that, with a large surface exposed to the air, the low temperature interferes considerably with the quantity of the cream. The conditions are thus exactly reversed to convenience, for when the most space could be given in winter time, it is not desirable to

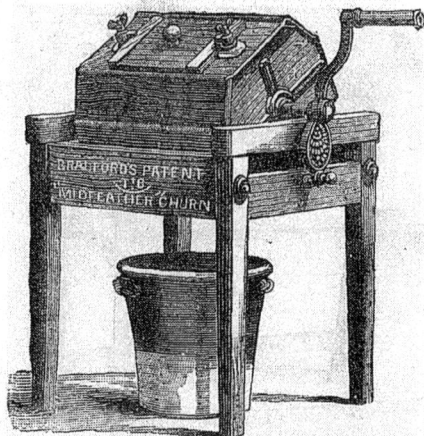

MIDFEATHER-CHURN.

make use of shallow pans, it being vitally essential to retain the natural heat of the milk as long as possible. To ensure this, the pan holding the milk is by some managers put into another containing hot water, which assists the rising of the cream, and renders the use of a stove unnecessary. This may be dispensed with, and the same result attained by putting about a cupful of boiling water in the bottom of each pan, when the weather is very severe.

70. **THE CHURN.**—Churns are of various sizes, from ten to a hundred gallons when worked by the hand, or double that size in large dairies which are worked by the aid of a small horse-gin. The old-fashioned implement called the *plunge-churn* is still extensively used, it being considered to act more efficiently than any other,

though it is very tedious and laborious in its operation, acting by means of a long handle inserted in a closed vessel, with a circular flat bottom; but this has now very generally given place to the *barrel-churn*, which is both convenient and suitable in every way, and when mounted on patent axles is everything that can be required. These axles consist of two small wheels set in a frame,

ALWAY'S TIN BARREL-CHURN.

and fastened one at each side of the churn-stand. The churn, on being lifted on to the stand, rests in the centre of these wheels, which revolve when the churn is driven, and thus materially lessen the friction, rendering the process of churning much less laborious. Where a large quantity of cream is churned, a horse-gear is attached to the churn, and a pony or horse set to work it, which is easily managed, and the animal put into the desired pace, so that after a while he will perform the operation without occasion for the slightest looking after.

In the making of butter a good deal necessarily depends upon the churn that is used, and it becomes highly necessary to have as good a one as possible, and of a kind the best adapted to the quantity of butter that is usually aimed at being turned out, so as neither to be too large nor too small. A good deal of labour is often thrown away in the process of churning butter by inexperienced people, from the condition of the temperature not being

OSCILLATING-CHURN

taken sufficiently into account, either of the atmosphere, the cream, or the churn, which may be respectively warmer, or colder, at certain times than at others, and the blame of butter coming slowly is sometimes put upon the churn, which is often entirely undeserved.

Various new churns have put in a claim, of late years, to be considered by dairymen, which possess some peculiar characteristic or other, some of which base their claims to recommendation upon producing butter quickly. This, however, is but of slight advantage when large quantities are made by the aid of horse-gear, as has been suggested; and by making it too quickly, a loss both in quantity and quality is commonly entailed. By over-heating any churn, butter may be made to come quickly, but a reasonable time should always be bestowed upon the operation, which may be reckoned at from twenty minutes to half-an-hour.

Another arrangement of churn, upon the principle of the band-churn, is for the barrel to be vertical, and worked by a foot-board. A man stands with each foot upon the treadle-boards, and by alternately throwing his weight on each flap, he draws down the cord on each side which is attached to the axle and fans, which turn backwards and forwards in the upright barrel-churn. The contrivance is simple and efficacious, but is seldom made use of. Another churn is in the form of a cradle, which can be easily swung to and fro, while a description has been given of a churn upon the same principle, made after the manner of a rocking-horse, upon which a child is put astride, and is thus taught to combine business with pleasure, making the butter while amusing himself.

The churn on page 58 represents a tin barrel-churn upon stand.

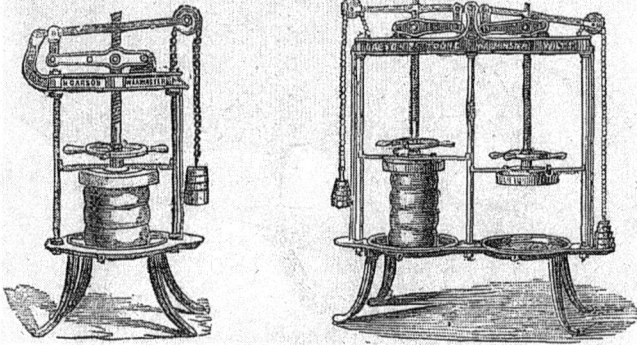

71. CHEESE-PRESSES.—Cheese-presses are made in various forms and weights, proportioned to the size of the cheese which is turned out, and vary from 5 cwt. to a ton. They are generally raised by a block and tackle, but some of them are upon the principle of the lever. Another very common form is of a simple arrangement, consisting of a movable beam fixed by a pivot in an upright post, and having hooked on at the other end a weight which presses on the cheese-vats underneath. This is generally used in turning out small cheeses, when so great pressure is not required. Another is made of iron, in a frame, and consists of a screw which is turned by a winch, the pressure of which can be regulated with greater certainty than by any fixed weight.

72. SINGLE CHEESE-PRESS.—The same is also made by Messrs. Carson and Toone, of Warminster, upon a double and treble principle, and cheese is pressed by various methods and

contrivances, some of them indeed of a very rough and makeshift order, which it is never worth while having recourse to, as dairy implements are now to be bought so cheaply, that will perform their allotted tasks with precision and despatch.

73. THE UTENSILS, however, as well as the fitting-up of the dairy, must be regulated by the nature of the business aimed at—whether the making of butter or cheese—and by the scale of operations which it is intended to set in motion. These must necessarily be considerations of primary importance, based upon the various capabilities of farm and situation; the object of everyone naturally being to secure as much profit as possible.

There are, however, some arrangements, the advantage of which appeal alike to all, a few of which we shall briefly mention,

Shelving of Stone or Slate.—Stone or slate shelves are better than wood, as being more easily cleansed; but better still, a stone or slate table should occupy the centre of the milk-house for the basins to stand on, so that they may be surrounded by fresh air equally, which can never be the case when placed in out-of-the-way corners, and along the sides of the wall.

The table should be water-tight, and, in the opinion of some, furnished with a water-tight ledge, so that cold or warm water may be thrown around the milk-basins when required. Of course the use of water for a definite purpose must not be confounded with unnecessary water standing about.

To others again, wood, as well as lead or zinc basins, are objectionable—the two latter for the sufficient objection, because they are liable to corrosion, or decomposition from the action of the acid contained in the milk, and the former from the difficulty of keeping the basins clean; but this latter will depend very much upon the dairymaid; from the use of which the advantages we have named may be secured.

74. DAIRYMAID'S DUTIES.—Everything will depend upon an efficient dairymaid, and her duties are pretty well defined in the hints upon management we have given in the foregoing, extreme cleanliness being the first essential.

Spilt milk should not be allowed to remain on the floors, tables, or shelves, a single minute longer than can be helped, and she ought to be unsparing in the use of plenty of water—cold in summer and warm in winter—and keep her dishes, and everything else, scrupulously clean.

A little common washing-soda, dissolved in water, will be found very useful in destroying any taint of sourness the various utensils may have acquired, which, if not removed, is apt to cause the milk to become sour before it would do so naturally.

Neither vegetables nor animal food should ever be admitted into the dairy; yet how commonly is it seen in small private dairies that the larder is united with it, and sometimes even raw meat placed in it on account of *its coolness for the meat!* By right, not even the cream-jars should be admitted. Cloths dipped in a solution of chloride of lime, and hung up on cords stretched out from side to side of the dairy, is a good mode of purifying the atmosphere.

75. DAIRIES IN TOWN.—Some very interesting general particulars relative to dairies in town were narrated by Mr. Morton in the *Journal of the Royal Agricultural Society*, who, in referring to results of Lodge Farm, Barking, where, from certain causes, the cost of each cow per week was far too high for the produce they yielded, speaks of the London dairies. As everything is done there upon business system and routine, we must give it as it is furnished, though, as it were, one part of the subject will run into another.

" Very little litter or other bedding is used. I have been over large suburban cow-sheds where there is none whatever used. The cows stand so close to each other that they cannot get across, and thus the dung and urine fall from them into the gutter behind them, from which it is cleared twice or thrice a day, and the lair—an earthen floor—is thus kept dry. At the Lodge Farm we have used sawdust. At present, 8 cwt. is the daily allowance in two sheds containing 85 cows, and there were exactly 21 tons of dung removed from these two sheds last week, being 3 tons daily. Most of the urine runs into a tank, only a portion of it being retained in the litter that is used. Two or three bushels of sawdust are, in the first place, put under every cow, and thereafter one bushel daily is sufficient, as much being daily taken away as fast as it gets soiled. The quantities amount to about 11 lbs. per cow added, and 80 lbs. of dung per cow taken, so that we collect about 70 lbs. per diem of the actual fæces of the animal. I may on this refer to a letter I received twelve years ago from Mr. Telfer, of the Canning Park Farm, near Ayr, who kept 48 of the small Ayrshire cows for a butter-dairy. He found that these cows yielded 60 lbs. of dung and 18 lbs. of urine every 24 hours. Taking their smaller size into account, this agrees very fairly with our experience at Lodge Farm. He adds that the cows yielding most milk, at the same time yielded the most dung and urine, which is not surprising, seeing that these are, in fact, the *débris* of a manufacture, and must be greater, or less, according to the quantity of raw material which passes through the machine. Mr. Telfer's cows lay on a cocoa-nut matting, their dung and urine falling into an accurately-made gutter, which was cleaned out perfectly by a single draw of a drag, made to fit the groove. In London cow-houses the rough causewayed floors are cleaned out with besom and spade into a dung-pit, which the sanitary inspector requires to be emptied at intervals, and the gutters in well-managed houses are washed down from the pail. Before referring to the produce of the cow-house, and to the quality and quantity of the milk obtained in it, it is proper very shortly to insist on the essential need of cleanliness. This, though especially required in the dairy, is desirable everywhere. The cow, like all other animals, is the happier and more healthy for it. The dairy vessels must, of course, be clean; the pails must be scoured and rinsed after every milking. The milk is poured from them through a strainer at once into the can or 'churn,' which stands ready to receive it at the cow-house door; and in a suburban farm it is at once lifted into the spring-van, which takes it directly up to town. Or in the case of a farm farther afield, the churn is placed to stand in water, and its contents are cooled down before being sent away. These churns must be scalded and rinsed after being emptied at the dealers'; and when returned to the farm they must be again scoured, and scalded, and rinsed, before being used. Having these, and providing as rapid a transmission as possible, the consumer will receive the milk at its very best."

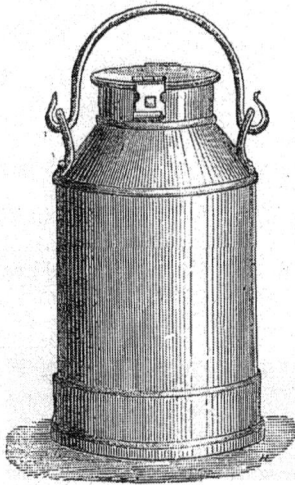

CREAM-BOTTLE.

CHAPTER VI.

MILKING.

Yield—Difference in Milk of Different Animals—Average Yield of Milk of a good Cow—On the Milking of Cows—Skimming, and the Treatment of Milk in Summer and Winter.

76. YIELD.—We will still further follow Mr. Morton in his account of the milk produce of the London dairies. It is to be remarked that, in almost all instances where individual care and attention is given by the owner, or some other really good and conscientious manager, the yield of the cows is always much greater than in others where only average interest is taken, and average pains-taking only is given.

"The quality of the milk depends upon the cow and the treatment of her, to which we have been referring. The milk of every cow has its own natural standard of quality, but taking the case of each apart, her milk is rich or poor—first, according to her nearness to the time she calved; and secondly, according to the quality of her food. The milk of a big, ordinary cow, bought half fat for a London cow-house, will throw up 14 to 16 per cent. of cream in three hours in the lactometer during the first few weeks after calving; the same cow similarly fed will not yield much more than half so good a quality when, after six or eight months' milking, she is rapidly diminishing her quantity. At an equal age, however, at the pail, the London cow, fed so as, if possible, to maintain or in-crease her flesh, will yield a richer milk than a country-fed cow which is being milked at grass. The way to keep a uniform quality when, as in London, a great part of the food (grains and hay) is constant throughout the year, is to keep buying in fresh cows in pretty constant numbers throughout the year. But except in the poorer districts, where the demand for milk does not vary throughout the year, this is not commonly done. A London cow-shed in the West-end, for example, is full only during the spring and summer months, when London is full; and as it is then that a richer milk is wanted for the sake of the cream which is required at 'good houses' during the season, that is the proper time to buy in freshly-calved cows. At many small cow-houses which I visited two years ago I was told that eleven, and even twelve quarts a-day are obtained on an average throughout the year; that is to say, a house of 10 stalls always full will yield 10 × 365 × 11 quarts of milk per annum, which is equal to 40,150 quarts, or 1,000 gallons per stall. If, as is possible, these cows are changed every eight months on an average, then 10,000 gallons is the quantity yielded by 15 cows during the eight months after calving before they are sold; each cow, therefore, yields 666 gallons in its eight months' milking. This, though a large quantity, is not incredible. In the case of the Frocester Court Dairy (Gloucestershire), of which a full account has been given in the *Bath and West of England Journal*, Mr. Harrison found that, of his 114 cows, 8 in the first year of milking (calving at two-and-a-quarter years old) yielded 317 gallons per annum; 15—also in their first year, but brought to the pail at three years— yielded 472 gallons; 14, in their second year, averaged 535 gallons; 15, in their third year, averaged 616 gallons; 20, in their fourth year, made 665 gallons apiece; 18, in their fifth year, yielded 635 gallons; 9, in their sixth year, made 708 gallons; 15 aged cows averaged 651 gallons apiece. These figures, however, give only an approximation to the truth, if they be taken to indicate the average yield of a cow at different ages, for doubtless, in a large herd like that of Frocester Court, the bad milkers, which would keep down the average of the first or second year, would be culled out, so that only the better cows would remain. It is cows in their third, fourth, fifth, and sixth year of milking, which are found in London dairies, and such cows at Frocester, depastured in the summer, yielded from 650 to 700 gallons of milk apiece per annum. They were, however, milked ten months, whereas the London cow is got rid of after eight months' milking in the case I have supposed. But the quantity of eleven, or twelve quarts a-day, which is the extreme report of some of the smaller cowkeepers, does not seem, on a comparison with Frocester, so incredible. On the other hand, if you consult the larger cowkeepers, supplying dealers who come and milk the cows, paying for what they take away, they will tell you that the average yield does not exceed nine, or nine-and-a-half quarts a-day to every stall. It is plain that, where cows are kept on till their daily yield is five quarts or less, in order to get fattened before sale, the average must be less than where the cow is got rid of sooner, and a greater loss submitted to upon her sale. On Lord Granville's farm at Golder's Green, Mr. Pauler, his lordship's agent, has told me that £3,900 was received one year for the milk of 100 stalls; in another year the sum received was £4,300 from 108 stalls constantly occupied; and in a third, £4,900 was received from 120 stalls. This at 1s. 10d. per eight quarts, which was the price received, amounts to 851, 868, and 891 imperial gallons per stall per annum, or 9¼, 9½, and 9¾ quarts respectively per cow per diem. This is where about 150 cows were purchased and sold every year, at a loss varying from £3 to £4 a-head, to keep 100 stalls constantly full. The cows were thus kept upon an average eight months each, and two-thirds only of the above quantities, 568, 587, and 594 gallons, are all that was taken from each cow during the eight months it was kept. I was informed that 89,236 imperial gallons were obtained in one year upon Colonel Talbot's farm at Sudbury from 80 stalls. The cows were sold earlier than at Golder's Green, not being kept longer on the average than six months, 153 having been sold and bought to keep 80 stalls full. In this case no less than 1,115 gallons was obtained per stall per annum, or fully twelve quarts per stall per diem. The cow here yielded 560 gallons in little more than six months, which is an enormous quantity for the average of so large a number as 80.'

It will be seen from the foregoing that it is the common practice in the London dairies to sell off the cows—to make a rule of so doing—when their milk begins to fall off, not after years, but in the current year; everything is sacrificed to the yield of milk, which is even forced at the risk of injury to the cow's constitution; and when she has done her utmost, she is sold, either to the butcher, or to anybody else who may happen to want a cow, and is willing to purchase her. Persons unacquainted with the extent to which this system is carried, have hesitated to buy a good cow from a London dairyman when wanting one for milking, thinking, very naturally, there must be some grave fault with an animal so disposed of, when the only object in getting rid of her often is that the natural rest enjoined by nature's laws cannot be afforded to be given to her in the London cow-shed, which the animal must obtain elsewhere.

This is quite contrary to country practice, where, in almost all cases, if a good cow is obtained, the owner does not want to part with her; but, appreciating the animal at its due worth, often refuses a good price for her. And this, of course, is as it should be; for in the country the breeding of calves is a very important part in dairy management, and calves are never wanted in the London cow-shed; it is a different line of business, and there everything is sacrificed for the milk. To those who have never had any experience of this system, the thoroughness with which everything is subordinated to this point is very remarkable, and the plan pursued will account for the high average per cow that is made to be given in the yield of milk from a certain number.

77. AVERAGE YIELD OF MILK OF A GOOD COW.—During the months out of one year a good cow is in milk, she will yield about 600 gallons. Many highly-kept cows will give more; but as a great many fairly-kept animals will produce less, it is not safe to calculate upon a larger average, and this result depends very much upon the species. Alderney cows give a much smaller amount of milk than most other sorts; but as much butter can often be made from a smaller yield of this description of animal, on account of its greater richness, as from a larger supply of the lacteal fluid; and this causes the Alderney cow to be in favour with private gentlemen, though they are not supposed to answer the purpose of a dairy-farmer so well. 600 gallons of milk at eightpence per gallon would amount to £20; but as sometimes as much as a shilling per gallon can be obtained by persons who are favourably situated for disposing of their milk produce, the large sum of £30 per cow can be got from the sale of her milk; but these results are not to be obtained unless the animals are liberally supplied with food of the best description, varied with brewers' and distillers'

grains, bean and Indian meal, &c., the method of feeding being kept up with the greatest regularity. An average of 650 gallons is commonly put; but we have in this instance put the figures at 600 for the sake of round numbers.

78. ON THE MILKING OF COWS.—The operation of milking the cows is, unfortunately, often conducted in a very slovenly manner upon some farms, and that attention is not paid to minute cleanliness which ought to prevail during the operation.

Many attempts have been made to milk cows by machinery, and some few years back the American " Cow-milker " was sold largely to cowkeepers and others, who hoped to get a useful contrivance of this sort, but nothing has yet been found which answers the purpose so well as hand-milking. The udder and teats of the cow frequently having particles of dirt adhering to them, which in the course of milking are apt to fall into the pail, they should be well brushed with the hand before commencing to milk; and if the dirt is soft or wet, they should be washed in tepid water. The washing should be avoided if possible, as sometimes the cold is apt to strike the cow, and dry wiping is the safest. Neglect of precaution often causes milk which would otherwise remain perfectly fresh to become tainted, and loss is sustained thereby.

The utmost care should be taken to drain the cow's udder well, or, as it is called, " drip the strippings " from her; for not only is this the richest part of the milk, but neglect of this important particular is apt to cause the cow to become dry. The operation should be performed as quickly as possible, without alarming or causing inconvenience to the cow. Young cows are often very timid and nervous, and from this cause are apt to misbehave themselves. Milkmaids are generally found to take more pains with the animals than men, when this is the case, as it is sometimes not unusual to see men throw the milking-stool at a cow which is not so tractable as it might be; and many a good animal has thus been spoiled by bad treatment. Pain, fear, or nervous excitement is highly injurious to cows, and, in young animals especially, tends to check the secretion of milk.

Many dairymen make a point of feeding their cows during the operation of milking, to put them in good humour; and the whole performance is done in such an orderly manner, that the milk-pail, instead of being a dreaded object by the animals, is the signal for so much enjoyment and gratification.

Everything that tends to ruffle the cow while she is being milked, should be avoided, and she should be kept as quiet as possible; and by good management this task may be made an easy and pleasant one, both to the cow and the milker, if considerate and gentle treatment is adopted. The animals quickly appreciate kindness, and can soon be made to learn what is expected from them.

A good deal of difficulty may be spared in anticipation, by considerate treatment to young heifers which are about to come into milk. They should be daily handled and petted, and made acquainted with the person whose task it will be to milk them. By giving them morsels of choice food, and allowing them to accompany the cows which are milked, they may soon be rendered docile.

Some young heifers are very wild when they are first milked, and by accustoming them to have their legs groomed and their udders handled, they will gradually be got into the way, and many a one which would otherwise have turned out a "kicker," has proved a docile animal enough when the time has come round for her to be milked, owing to the precautions and the trouble which have been taken with her beforehand. The traditionary cow which gave the good pail of milk and then kicked it over, was doubtless one which had to deal with a bad-tempered milker while a heifer; and the great point to be observed is, never to give the animal pain, or excite her fears. Many heifers are annually spoiled by hasty and injudicious treatment in "breaking them to the pail," which can best be done by kindness, and by humouring them. The punishment often adminis-tered to an animal in the shape of kicks and blows at the time of milking, is naturally calculated to make it hate the sight of the pail, and to stir up appre-hension, when there ought to be no occasion for it.

Where cows have been kindly treated, it is no uncommon thing to see them answer to their names when called, and come up at a trot to be milked, when the milker has ingratiated herself in their favour; but this sight, it must be admitted, is more frequently seen abroad than in England, and cows are capable of being rendered very docile by kind and judicious treatment.

79. SKIMMING, AND THE TREATMENT OF MILK IN SUM-MER AND WINTER.—Milk is generally skimmed in England for the purpose of making butter, and there are one or two points about this operation which deserve mention.

In cold weather the cream does not rise so rapidly to the top of the dish as in warm, so that while it is usual in summer to skim it two or three times, it is skimmed as often as four times in winter, or continued till no more cream can be got from the milk. To perform this operation dexterously, as the cream adheres firmly to the sides of the pan, it should be separated from the edges by running an ivory or silver knife round it. The cream should then be carefully lifted with the "skimmer," which is generally per-forated with small holes, to prevent raising any of the milk with it. There is a method followed by a few, who have a plug in the bottom of their milk-pans, which they remove, and allow the milk to flow off, leaving the cream behind; but skimming is the ordinary practice followed. The length of time that the milk should stand depends a good deal upon the temperature. In warm weather, eight hours is the least, and about twelve hours the average; while in winter it will have to stand much longer. The cream is then placed in a "cream-pot," the most perfect kinds of which have a tap near the bottom, so as to draw off any thin, serous por-tions of milk which may chance to be there, which, if allowed to remain, act upon the cream and greatly deteriorate the quality of the butter. The contents of the cream-pot should be stirred every day with a wooden spoon, in order to prevent coagulation, until enough is collected to put into the churn. A common error pre-vails, that no butter can be first-class which is not made from fresh cream. The formation of butter only takes place when the cream has imbibed a certain degree of acidity, and no good butter can be made from cream that is not more than one day old.

Judgment and experience are the best safeguards to rely upon, as to the length of time cream should be allowed to stand, as its condition varies from altered circumstances. Cream that has been kept three or four days is in excellent condition for making butter in summer, but if the cows are fed on roots, or artificial grasses, or the herbage is coarse, then the sooner the cream is churned the better. The cream from every milking should be kept separate, till it becomes sour, and not mixed with sweet cream until the moment of churning.

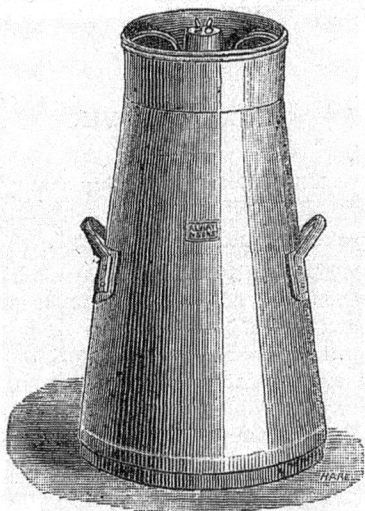

" ACME " CHURN.

MILK CARRIAGE.

CHAPTER VII.

MILK.

Properties of Cow's Milk—Adulteration—Whey Butter, Whey, &c.—Cream: Clotted or "Clouted" Cream—Skimmed Milk—Milk considered as an Aliment—Varieties of Food prepared from Milk—Markets for Milk—Transport of Milk—Cost of Production and Profits—Dairying in Flanders.

80. **PROPERTIES OF COW'S MILK.**—We have already spoken of the differences in milk of different animals and under different circumstances, and while in most cases its chemical proportions and properties will not particularly interest the general reader, a practical experiment to determine the butter and cheese-making properties in milk will doubtless be found interesting.

A very definite experiment which is recorded as having been made, published in Morton's Cyclopædia of Agriculture, illustrates this point in a very conclusive manner, the object being to determine the exact quantities of butter and cheese in the milk of each cow :—

" A weighed quantity of milk was taken from the noon's milking of each cow, and allowed to stand in separate glass vessels for forty-five hours. A portion of the *afterings* of all the cows, mixed, was also set apart, to determine the amount of butter and cheese in the last-drawn milk. When the cream had completely separated from the milk, a fine-pointed glass syphon—sufficiently wide in the bore to allow the milk to run through it, but not the cream—was introduced into the vessel, nearly touching the bottom. The air was then exhausted from the syphon, and the milk withdrawn into another vessel. The cream was weighed, and agitated in a glass tube until the butter came, which was then well washed with pure water, and repeated decantings until the water ran off colourless. The weight of the butter was then carefully ascertained; and the difference between it and the weight of the cream gave that of the butter-milk. The butter was then put in a minim tube, and melted at a low temperature, by immersing the tube in warm water. The remaining butter-milk and cheesy matter sank to the bottom on cooling, and the proportion, by bulk, was noted down.

"The skimmed milk was gently warmed to 90°, after adding a little acetic acid to make it curdle. The whey was separated from the curd by filtration and washing, and the latter then dried at a heat not exceeding 212°, until it ceased to lose weight. The weight of the dried curd (pure caseine), when deducted from that of the milk, left, as a remainder, the weight of the whey. The following table shows the relative quantities of butter, caseine (cheese), and whey; the latter includes the butter-milk also :—

Per Cent.	Middle-sized, well-proportioned cow; colour, very dark brown.	Fifeshire breed, long body, broad behind and narrow before. Black.	Cross-breed from short-horn, very broad, square cow. Brown and white.	Fife breed—heavy body, wide chest. Black.	Angus breed—low, square, well-proportioned figure Black and white.	"Afterings" of the five cows.
Butter	4·318	4·209	2·900	3·079	4·700	10·102
Caseine (Cheese)	3·017	3·412	3·144	3·389	3·209	3·294
Whey, &c................	92·665	92·379	93·956	93·532	92·091	86·604
	100·000	100·000	100·000	100·000	100·000	100·000

"The large proportion of butter in the last-drawn milk is seen from the figures in the last column. It indicates the truth of the remark we once heard made by a dairy-farmer, that the profits of his business depended principally on the perfect performance of the operation of milking.

"The quantity of milk, daily, from each of these cows, during seven days in the month of July, was as follows :—

	Qts.	Qts.	Qts.	Qts.	Qts.
Daily	No. 1, 9⅗	No. 2, 12⅘	No. 3, 13½	No. 4, 10⅗	No. 5, 10¼
Weekly ...	No. 1, 68	No. 2, 89	No. 3, 96	No. 4, 75	No. 5, 72

"If we take the weight of a gallon of milk at 10 lbs. 3 oz., the weekly yield per cow, of butter, cheese (caseine), and whey, would be as follows :—

	No. 1.	No. 2.	No. 3.	No. 4.	No. 5.
Produce per cow in qts. and lbs.	68 qts.= 173³⁄₁₆ lbs.	89 qts.= 226¾ lbs.	96 qts.= 244½ lbs.	75 qts.= 191 lbs.	72 qts.= 183¾ lbs.
	lbs.	lbs.	lbs.	lbs.	lbs.
Butter.......................	7·479	9·540	7·09	5·881	8·620
Caseine	5·225	7·734	6·89	6·473	5·885
Whey, &c...................	160·496	209·476	230·52	178·646	168·895
Total............	173·200	226·750	244·50	191·000	183·400

"Of course, the caseine in this table does not represent the whole of the *cheese*

F

which the milk contained, because the process employed to extract it separated the butter entirely from it; besides, the cheesy matter was dried to the consistency of horn before being weighed. Common-milk cheese, however poor, as it is usually made, not only contains a little of the butter, but also a large proportion of water or wheyey matter. On the other hand, the quantity of butter given above is, no doubt, larger than could have been obtained by common churning. Still the table will serve to show correctly the comparative, as well as absolute, amount of pure butter and caseine contained in the milk of each cow.

"It will be seen, from these statements, that the money value of each cow would fluctuate according to the purpose for which she was kept; whether for milk, butter, or cheese. Calculating by the milk, at 6*d.* per gallon, the value of each cow, weekly, will stand thus:—

			s.	*d.*
No. 1.	17 gallons at 6*d.*	... ...	8	6
No. 2.	22¼ ,, ,,	... ...	11	1½
No. 3.	24 ,, ,,	... ...	12	0
No. 4.	18¾ ,, ,,	... ...	9	4½
No. 5.	18 ,, ,,	... ...	9	0

"Again, supposing the milk all to be churned, and sold as butter and butter-milk, the result would be as follows:—

		s.	*d.*	*s.*	*d.*
No. 1.	7·47 lbs. of butter at 10*d.*	6	2¾		
	16 gallons of butter-milk at 3*d.*	4	0	10	2¾
No. 2.	9·54 lbs. of butter at 10*d.*	7	11½		
	21¼ gallons of butter-milk at 3*d.*	3	3¾	11	3¼
No. 3.	7·09 lbs. of butter at 10*d.*	5	11		
	23¼ gallons of butter-milk at 3*d.*	5	9¾	11	8¾
No. 4.	5·88 lbs. of butter at 10*d.*	4	11		
	18 gallons of butter-milk at 3*d.*	4	6	9	5
No. 5.	8·62 lbs. of butter at 10*d.*	7	2¼		
	17 gallons of butter-milk at 3*d.*	4	3	11	5¼

"The reader will see, from these tables, that the cow No. 3, although giving six gallons of milk more than No. 5, and seven gallons more than No. 1 per week, is under both of them in butter; and were it not that the quantity of butter-milk is great, she would fall below them in profit too. Her milk is poor in butter and cheese, and there is reason to suspect that the quality of both is inferior also. To the inland dairy farmer, it is of the greatest consequence to get cows that yield rich milk, even although the quantity should not be so very great; for this reason, that the refuse, either of cheese or butter making, can be turned to little account in such localities."

We disagree with the opinion that little can be done as regards skim-milk, because calves can be brought up profitably on it, when butter is made from cream alone, the method of doing which is explained in another place.

81. **ADULTERATION.**—Since the Adulteration of Food Act has come into operation a great many persons have been fined for mixing water with milk, and care should be taken by those who sell milk in large quantities to have the cans properly sealed, so that

they cannot be tampered with on the journey towards their destination, whether it be by cart or railway.

The pump—the cow with the iron tail—has often been described as the most profitable animal of the whole herd ; but those days are now over for those who are dishonestly inclined. Clean water is not in itself so objectionable a form of adulteration as at one time was resorted to ; when, in order to counteract the poorness of quality communicated to the milk by too great an allowance of *aqua pura*, the intestines of animals were boiled up, and the liquor in which they had been cooked mixed with the milk.

82. **WHEY-BUTTER, WHEY,** &c.—The form in which the·produce of the dairy is put varies considerably in different counties and different districts. · Where whey-butter is made, it is usual to heat the whey in a set pan to 180°, and frequently stir it to prevent it from burning. A little sour butter-milk and white whey (thrustings, as the latter is called in some districts), in the proportion of 1 pint of the former and 2 quarts of the latter to 22 gallons of whey, are thrown in, upon which the cream immediately rises to the surface, and is skimmed off and put in a jar to sour or clot.

In a few hours after being placed in the jar, the thicker and more oily part of the cream rises to the top, and the thin wheyey matter is withdrawn by a spigot from below. In three or four days the cream is completely clotted, and ready for being churned, which is done in the usual manner. This is the method followed in Cheshire.

In Gloucestershire dairying, in autumn and winter, when the weather is cold, a small portion of the milk is heated and mixed with the other, so as to bring the whole up to the temperature of 85° before adding the rennet, and the milk allowed to remain for an hour without disturbance. During this time it is covered closely over with a woollen cloth, to exclude the cold air. By that time, if matters have proceeded properly, the curd will be completely formed, fit for being broken up, which is effected by passing a three-bladed knife or a coarse wire sieve gently downwards to the bottom of the tub.

When the curd has been cut through and divided as well as its suspension in the whey will allow, the whole is allowed to remain for ten minutes or so, undisturbed, to allow time for the broken curd to sink, so as to allow the whey to be baled off the top.

As soon as all the clear whey has been taken away, the curd, which is now more consolidated, is again broken, but more slowly than before, to avoid squeezing out any of the butter, which would not fail to ensue if the curd were cut too rapidly, or in a rough manner. The curd properly broken and reduced to an equal degree of firmness, it is allowed to settle for a short time, when more of the whey is removed and poured through a sieve, to retain any small particles of curd which may yet adhere to it ; and when the greater part of the whey has been removed in this way, the curd is separated into lumps and laid aside one upon the other in the bottom of a tub placed in a somewhat tilted position to allow the whey to drain away and be removed, and as soon as it ceases to drain the curd is ready for being placed in the vat, when the subsequent operations for making the cheese are commenced.

83. CLOTTED OR CLOUTED CREAM.—The method followed in Devonshire and other western counties for procuring "clotted cream" has been described as follows :—

The milk, while warm from the cow, is strained into either large shallow pans, well tinned, or earthen ones, holding from two to five gallons, in which should be a small quantity of cold water. This is thought to prevent the milk from burning, and to cause the cream to be more completely separated and thrown to the top.

The morning meal of milk stands till about the middle of the day; the evening meal until the next morning. The pans are now carried steadily to and placed over a clear slow fire; if of charcoal, or over a stove, the cream is not so apt to get an earthy or smoky taste, as when the milk is scalded over a turf or wood fire. The heat should be so managed as not to suffer the milk to boil, or, as they provincially term it, "to heave," as that would injure the cream. The criterion of its being sufficiently scalded is a very nice point; the earthen pan having its bottom much smaller than the top, allows this point to be more easily ascertained, because when the milk is sufficiently scalded the pan throws up the form of its bottom on the surface of the cream.

The brass pan, if almost as big at the bottom as at the top, gives no criterion to judge by but the appearance and texture of the surface of the cream, the wrinkles upon which become smaller and the texture somewhat leathery. In summer, it must be observed, the process of scalding ought to be quicker than in the winter, as in very hot weather, if the milk should be kept over too slow a fire, it would be apt to run, or curdle.

This process being finished, the pans are carefully returned to the dairy; and should it be the summer season, they are placed in the coolest situation, if on stone floors or slate benches the better; but should it be the winter season, the heat should rather be retained by putting a slight covering over the pans, as cooling too suddenly causes the cream to be thin, and consequently to yield less butter, the mode of making which is this: The cream should, in hot weather, be made into butter the next day, but in winter it is thought better to let the cream remain one day longer on the milk. The cream, being collected from the pans, is put into wooden bowls, which should be first rinsed with scalding, then with cold water. It is now briskly stirred round one way with a nicely-cleaned hand, which must also have been washed in hot, and then in cold water; for these alternate warm and cold ablutions of bowl and hand are not only for the sake of cleanliness, but to prevent the butter from sticking to either. The cream, being thus agitated, quickly assumes the consistence of butter; the milky part now readily separates, and, being poured off, the butter is washed and pressed in several cold waters; a little salt is added to season it, and then it is well beaten on a wooden trencher until the milky and watery parts are separated, when it is finally formed into prints for the markets.

The dairy-maids say that one-fourth more cream is obtained this way than by the ordinary method of skimming it off the milk.

84. SKIMMED MILK.—Skimmed milk, in a well-managed dairy, can be made a very important element of profit, though very often it is much overlooked and neglected. Both in the rearing of calves and in the feeding of pigs, skimmed milk can be made to play a very important part, of which we shall speak hereafter under the heading of each. Properly managed, it pays much better to con-

sume it in feeding stock than to sell it, though it is generally felt that the neighbouring poor cottagers ought to have the opportunity of purchasing enough at a low rate for their household requirements.

85. **MILK CONSIDERED AS AN ALIMENT.**—As an aliment, milk is considered a necessity for young children, and one of the most important articles of food, as it is easy of digestion, and yet contains within it those nourishing and sustaining principles which are so valuable in imparting health and strength to the growing frame in childhood; while, under certain conditions, it is equally useful to aged persons and invalids.

The demand for milk as an article of diet is daily increasing, more faith in its alimentary properties being generally entertained since the provisions of the Adulteration of Food Act are being strictly carried out in cases of adulteration.

86. **VARIETIES OF FOOD PREPARED FROM MILK.**—Besides butter and cheese, which are the two great staples prepared from milk, it enters very largely into the composition of various kinds of food consumed in the household, in the shape of custards, puddings, &c., so that it makes a very necessary adjunct to the daily food of the people. In every household, nearly, throughout the country, the milk-jug is in requisition at each morning and evening meal as an accompaniment to that " cup which cheers, yet not inebriates," and is both an important item of diet itself, as well as entering largely into the composition of others.

87. **MARKETS FOR MILK.**—There is literally a market for milk at one's very door where any amount of population exists, and there never is any difficulty in finding one, however isolated a dairy farm may be, either through means of the railway, or by horse and cart despatch, or by the two latter conjointly. Of course, where milk is sent to market at a distance, the cost of transport must be taken into account, and each item of expense connected with it carefully estimated.

A good market is a very important consideration, and nearness to one often means a very great accession of profit, and this factor in the general calculations ought to be carefully taken into account when renting a dairy farm.

88. **TRANSPORT OF MILK.**—The facilities which now exist for the disposal of milk at long distances through means of the railways are very great, which may be gathered from the statement made some time back by Mr. Brooks, of the London and North

Western Railway, to the Milk Committee of the Society of Arts, as
to the methods of charging and delivering milk between London
and Northampton.

"The milk is conveyed in cans, which are provided by the senders, in open
carriage-trucks; the carriages being well constructed in respect to springs, so far
as this can be done, in order to cause the milk to be as little shaken as possible
in the course of the journey. The charge for a distance not exceeding 100 miles
is 1½d. per imperial gallon, and when the distance exceeds 100 miles, 2d. per gallon.
When the great increase in the traffic first commenced, milk was sent up from
places as much as 180 or 200 miles distant, from districts near Huddersfield,
Macclesfield, &c., the greatest distance at which milk was then sent being 95
miles. The cans in use in England are much too large and too heavy to be
loaded and handled by one man, and it is a stipulation with the dealers that their
men shall assist the railway porters in unloading the trucks, the weight of a can
filled with milk being nearly 200 lbs. The French cans are about half the size
of those used here. The shape of the English cans, too, is against their being
closely packed, being broad at the bottom and tapering towards the top. The
French tins, on the contrary, from their cylindrical shape, can be packed with
greater economy of space. The French milk-trucks are very much like the
narrow-gauge sheep-trucks used in this country, with two floors, one above the
other, forming two tiers, in which a great number of cans can be packed, and
there is a good circulation of air around them. On the other hand, the French
cans are heavier per gallon of their contents than the English, and it is not likely
that the former will be adopted here. The trade has got to be such an important
one as to lead to the dispatch of special trains for this purpose, and the milk
is brought to the stations at specified times to meet them. One train arrives in
London at a quarter to twelve in the forenoon, for the afternoon supply of the
metropolis; and the second train arrives about half-past eight in the evening,
for the next morning's supply. During the time of the greatest scarcity of
milk, an arrangement was made for bringing cream from a distance so remote
as Carlisle, which was placed in small cans, much smaller in size than the French
milk-can, and carried suspended in the railway truck; but when it arrived in
London it was found that the cream was reduced almost to the consistency of
milk, and the trade was therefore abandoned."

Mr. Brooks, upon the occasion referred to, stated that those who made com-
plaints about the rates of carriage cannot have calculated the price per ton at
which the Company carry the milk, or they would have found that the milk,
including the weight of the cans, is carried a distance of 100 miles for 1s. per
cwt. When the milk-train arrives the dealers assist in unloading the vans, and
the milk is carried away in the dealers' own conveyances.

The consumption of milk being so great in London, a large trade
has sprung up there for supplying it to the public, and a considerable number of cows are fed for this purpose both in the suburbs
and London itself, to which we have previously copiously alluded.

89. COST OF PRODUCTION AND PROFITS.—Both the cost of
production and profits being relative matters, which depend upon a
certain number of contingent circumstances, no definite scale can
be furnished of either with reliable accuracy; but we refer the
reader to those instances that we have furnished, where certain
prices are quoted for milk that is sold, and also the cost of keep of
the cows. The latter is made to vary considerably by the amount

of artificial and stimulating food which is given, or otherwise; a fair conclusion from which may be drawn from the examples we have instanced.

The yield of the cow can be considerably increased by extra feeding, but, on the other hand, the enlarged expense must be taken into account; but, as we have previously shown, this can be done so as to ensure an extra rate of profit; this again depending upon the conveniences, or otherwise, for the supply of artificial food.

It is just possible that, in a purely pastoral district, where only grazing cattle are fed, and the difficulties of transit and cost of carriage are great, merely feeding the animals upon grass and hay, supplemented by light portable artificial food, would pay in the long run, as well as the more complicated systems of feeding which have been recommended; but of course judgment must be used in each varying condition, and the relative cost and profit will naturally be subservient to them.

90. DAIRYING IN FLANDERS.—In Flanders cows are chiefly kept for the sake of the milk and the manure, oxen seldom being used there for tilling the land, which is an important consideration in some countries. The cow-stalls are littered two or three times a day with rye or oat straw, at the rate of twelve to fourteen pounds per head.

There is a very large variety in Flemish cattle in the north of Flanders, where pasturage is more abundant, but coarser than in the south; they are much the heaviest, though dairy cows are not superior.

A good many cattle are imported from Brabant, those which come from the Kempen being larger than the Flemish oxen. Cows are generally brought on the pastures about May, and remain on them till October or November, many farmers keeping the cows always in the same pastures, as they graze closer than the oxen, which causes the grass to become softer.

The best pastures are to be met with in the north of West Flanders, and chiefly in the district of Dixmuiden.

The cows are milked three times a-day, in the best season of the year, into a brass or wooden pail. In the cow-stalls a brass can of about four gallons stands, and upon it is placed a sieve, into which the milk is thrown out of the pail. When full, the can is immediately brought to the cellar, and there the milk is again poured through a horsehair sieve into the milk-tubs, which stand on a platform built on the ground for the purpose.

It is customary to churn the milk in summer after twenty-four hours, but in winter three days are allowed to expire, when the milk is poured into the tub that stands in the cellar, and if it then does not become sour, they place a can with warm water in the tub to accelerate it.

Those who sell fresh milk skim it some hours after it has been placed in the cellar, and pour the cream into a tub till it is churned. The churning generally lasts one-and-a-half or two hours, all the vessels being scoured and scalded out as soon as they are empty, the utmost cleanliness being observed, while the cellar is kept cool.

Two-and-a-half gallons of good sweet milk will produce a pound of butter

When the butter is taken out it is kneaded in a wooden dish with a wooden spoon, after which it is put in an earthen basin and covered with water. In an hour's time it is salted, two handfuls of salt being worked into every seven or eight pounds of butter; after which it is worked again with the wooden spoon, to free it from any remaining milk, and the butter is made up ready for market. As soon as the butter comes out of the churn it is cut through with a hair knife in opposite directions, the operation being known by the name of *combing*.

The months of May and September are considered to be the proper ones to make up butter for winter. It is then worked up a second time by kneading it in a wooden tub, after which more salt is added to it, when it is put in casks and vessels and kept under pickle.

NOSE RING.

BUTTER WORKER.

CHAPTER VIII.

BUTTER.

Utensils, &c.—Process of Making Butter—Precautions for ensuring Good Butter—
Feeding for Butter-Making in Winter—Varieties of Butter—Butter made from
whole Milk—Adulteration—Imitations of Butter—Butter as Food—Markets—
Importations—American Factories for Butter Manufacture—Salting—Plan of
Working the Butter—Advantages of Butter Factories—Skim Cheese—Results
obtained at the Butter Factories—Labour, &c.—Cost of Production and Profits.

91. **UTENSILS**, &c.—The milk is skimmed by a shallow pierced
tin ladle, which lifts off the cream from the surface of the milk-pan
when it has risen ; it is then stored in stone jars, or " cream-pots,"
until enough has been accumulated to place in the churn.

A great variety of forms exists with respect to the shape of the churn ; the
barrel-churn, perhaps, being that most commonly used, the barrel being turned
by a handle, and the milk, lifted by dash-boards extending radically inwards from
the sides, is shaken by them in the revolving motion. The price of these churns
varies according to size, and they may be purchased from £2 upwards. One with
a barrel eighteen inches long, and eighteen inches in its largest diameter, with two
beaters projecting inside and attached to the staves, would churn the milk of six
cows and cost about the sum named.

When a smaller quantity of milk is churned, where only a cow or two is kept,
and the dairy operations are upon quite a small scale the plunge-churn is mostly
used. It is also resorted to in some large dairies where milk, and not only cream,
is churned, being then made large enough to hold sixty gallons or more, the
plunger being worked by a crank movement lifting a lever, to the end of which
the churn-staff is attached, and by this means worked up and down. Many
other kinds of churns exist, but the two mentioned include the principle upon
which nearly all the churns in common use in England act.

The common box-churn is a rectangular wooden box, about seventeen inches
by twelve long, and sixteen inches deep, bevelled below, so as there to offer an
octagonal section in the vertical plane in which the beaters revolve. A revolving

frame of flat wooden beaters is contained inside, and when large enough to make ten pounds of butter, the cost of this churn would also be about £2.

92. **PROCESS OF MAKING BUTTER.**—Numerous experiments have been made, to show that the quantity of cream has nothing to do with the time of churning, providing the proportion of agitating surface is made to suit the capacity of the churn. Thus in some churns it has taken sixty-one minutes to produce a quantity of butter that has been made in another in twenty-five minutes.

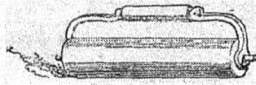

BUTTER ROLLER.

In the most effective churns there are two sets of beaters, which are made to revolve in different directions, thus bringing a large quantity of working surface into action.

After the churning has been completed, the butter is either made up and put into rolls, or forced into moulds, or made up into the most suitable form for market that is best liked in the district where it is made. In London, especially, a good deal of butter is made

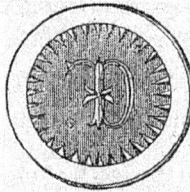

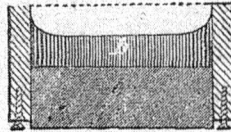

BUTTER PRINTING CYLINDER.

up in small quantities that are denominated "pats," a great part of this being disposed of to hotel proprietors and coffee-house keepers, to suit the wants of their various customers.

A great many persons make a practice of washing their butter as it comes from the churn, with the object of extracting all the milk which may remain in it, which they ascertain to have done when the water comes from it pure. The practice is not, however, a good one, and where a dairy is managed in a first-class manner is never followed, for experience shows that butter retains its sweetness considerably longer in those instances where water has not been used in making it up. If, when the butter is taken out of the churn, it is well worked with the hand in an effectual manner, the milk can be pressed out. Too much squeezing or working, however, must be avoided, as the butter is apt to become tough, and by pressing a cloth repeatedly upon it, any particles of milk remaining on it will be absorbed. There is always a full supply of inferior butter upon the market, and a comparative scarcity of a first-rate article, and it is not at all an un-

BUTTER WORKER.

common circumstance for one farmer in a district to feel the effects of glutted markets and low prices, while his next neighbour can sell all the butter he makes very readily at comparatively high prices. The secret of this is, that the

BUTTER WORKER.

article in the case of one is turned out of first-rate quality, while in that of the other it is inferior, the latter result being often due to want of sufficient care in its manipulation.

93. **PRECAUTIONS FOR ENSURING GOOD BUTTER.**—There are one or two precautions for ensuring the making of good butter, which should be ever taken. First: Good ventilation in the dairy. Nothing taints milk so soon as damp and confined air. In summer coolness is necessary; for if the dairy is too warm the milk thickens at once, and the butter is so soft that it is a work of considerable difficulty to make it up and prepare it for market. In winter, if too cold, the cream does not rise well, and there will be, in consequence, a certain amount of loss. Second: Strict attention to cleanliness, and seeing that every article used in the manufacture of the butter is thoroughly clean, dry, and sweet. Third: Not allowing the milk to stand too long in the pans. Allowing it to stand too long is a very common source of mischief, the butter losing the fresh sweet taste it would otherwise have, and often acquiring a taint. Fourth:

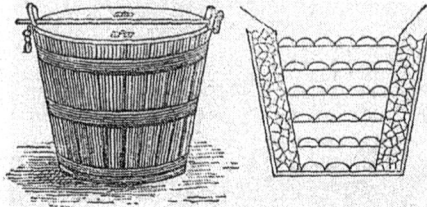

AN ICE BUTTER-TUB.

Churning often; those who attach considerable importance to this making a point of churning three times a-week. Fifth: During the winter months retaining the natural heat of the milk by the best methods as long as possible. Some dairy managers, to ensure this, put the pan containing the milk in another holding warm water; but if the adequate warmth can be communicated to the dairy the better. Sixth: During very hot weather in summer—particularly on very rich pastures—butter which, under ordinary conditions, is of the best quality, gets spongy, and wants texture, so that it is next to impossible to impart firmness to the mass. This can be remedied by trimming a slate to fit the top of the butter-firkin without touching the firkin itself, a layer of salt being placed between the slate and the butter to prevent contact, on the top of which is placed a heavy weight—say half-a-hundredweight.

In forty-eight hours the water will be forced out of the butter, and the consistence of it will be all that is required.

94. **FEEDING FOR BUTTER-MAKING IN WINTER.**—In order to ensure good butter in winter, it would be the best course in this place to revert again to the question of feeding, upon which everything will depend, and the chief point of importance is to give the cows the best and most appropriate food that can be given to them under varying conditions.

The different plans in connection with "soiling," or house-feeding, that are followed by the London cowkeepers which we have instanced, country dairymen can, without doubt, take some good hints from, even those who have plenty of pasture on which their cows are turned for, at all events, five months of the year out of twelve, during which period they have but comparatively little trouble, and good butter may be relied on, unless the herbage is rank or rushy, or some noxious weed abounds. But during the winter and early spring months, however, the case is different, and the country dairyman is in much the same position as his brother in town; and now comes the test of management. As cows during the remaining seven months of the year are mostly fed upon roots—that is to say, the bulkier portion of the food consists of roots—they are apt more or less to communicate a disagreeable flavour to the butter, which often not only lowers its commercial value, but at times renders it very difficult to be disposed of at all. Turnips may be freely given to cows whose milk is intended for butter, if concentrated food is mixed with them, the method of doing which is described elsewhere, and the butter be little inferior to that produced in summer.

The roots used must, however, be sound, and not have heated in the pits, turnips being considered productive of butter. When boiled food is given morning and evening, meal can be mixed with it very advantageously, so that they get the most benefit from it; and where boiled food is not given, the concentrated food can be thrown into a large tub, and hot water poured over it, the steam being confined in it by a cloth or cover. By this means the food receives a certain amount of cooking, and the cows eat it with relish. Some do not take this trouble, but sprinkle the mixture, or whatever it may be, either crushed oats, a mixture of meals, bran and oil-cake, Indian or palm-nut meals, &c., &c., over each animal's allowance of roots. When this has been practised for some little time, the cows themselves will remind their attendants of any omission on this point occasionally, for some of them will not touch their turnips or mangolds till the meal has been sprinkled over them.

A difference of opinion exists as to the comparative merits of mangolds and turnips as food for cows by different dairymen. By some, mangolds are supposed to make a better quality of butter than that got from turnips. Others, on the contrary, say that while the risk is run of butter tasting of turnips, the somewhat peculiar and slightly bitter taste which is communicated to butter by mangolds is equally objectionable to many as "turnip butter," and is more difficult to be got rid of. As mangolds keep so well, they will always be used as food for cows in winter; yet they are not so good for making butter as turnips, the milk being poor and thin, and the cream not rich in butter. Where mangolds are freely used, then it is imperative that richer food be given in addition, not only for the sake of improving the quality of the milk, but also for the purpose of keeping the animals in health; for the acrid juices which are found in mangolds even late in the season have, without a mixture of other food of a retentive character, a purgative effect upon them, which reduces their condition very much.

Turnips and hay alone are sometimes given, it is said, without injury to the taste of the butter, the hay qualifying the effect of the turnips; yet the same with mangolds is not sufficient, and if the health of the cows and the yield of their milk is studied, to mangolds and hay must be added a portion of meal or other nutritious food. From four to eight pounds of meal a-day is a sufficient allowance for each cow, and this ensures its health, enriches the milk, and increases its quantity.

95. **VARIETIES OF BUTTER.**—The varieties of butter are chiefly comprehended under the heading of two main divisions, salt and fresh—the latter being intended for immediate sale and consumption, and the former put away in tubs or barrels, with an addition of salt to preserve it sweet, for store use; but butter is also made from unskimmed milk, and not cream alone.

96. **BUTTER MADE FROM WHOLE MILK.**—The process of making butter from the whole milk, and not from cream alone, which is practised in some of the dairies of the West of Scotland, has been thus described :—"The milk, when drawn from the cow, is placed in the coolers on the floor of a clean, cool, and well-aired milk-house, from twelve to twenty-four hours, till it has cooled to the temperature of the milk-house itself, and the cream has risen to the surface. These coolers are next emptied, while the milk is yet free from acidity, into a clean, well-scalded vat, of size to contain the whole milking, or two milkings, if both are sufficiently cooled, where it remains till churned. If another milking, or meal of milk, be ready before that which has begun to become sour, that second meal may be put into the same vat; but if the first has soured, or is approaching to acidity, before the second quantity has completely cooled, any further admixture would lead to fermentation and injure the milk. It is necessary that the whole milk become sour before it is churned, but the whole of it must become so of its own accord, and be by no means forced into acidity by any mixture of sour milk with that which is sweet. The utmost care should, however, be taken not to allow the coagulum, or curd, of the milk in the stand-vat to be broken till the milk is about to be churned. If it be not agitated, or the 'lapper' (as it is termed in dairy parlance) broken, till it is turned into the churn, it may stand from a day to a week without injury. If these rules be attended to, the butter will be rich, sound, and well-flavoured, and the buttermilk will have a pleasant, palatable, acid taste; but wherever fermentation has been excited, or the lapper broken, and the milk run into curds and whey, the fermentation so begun will continue in the butter-milk after that operation, and will become acrid and unwholesome. When duly prepared and manufactured, the milk will be the better with a fifth or a fourth part of water mixed into it, than milk that has been fermented before being churned would be without a drop of water mixed with it."

97. **BUTTER-MILK.**—Butter-milk is an article of food largely consumed in Scotland, and the foregoing method of churning the milk results in a great quantity of butter-milk being produced. In England the skimmed milk is more or less appreciated, according to locality and circumstances, and is in that form consumed in those districts where the butter is made from cream only.

98. **ADULTERATION.**—Perhaps at one time there was no article of food so largely adulterated as butter. The adulteration chiefly consisted of melted fat, which was bought from the butchers in the form of suet and otherwise, and was mixed with the butter after undergoing a certain process. Large quantities of this fat used to be sent abroad, and came back to us in the form of tub butter, which, although not injurious to health, defrauded the purchaser by substituting an inferior article and a different one for that he thought he was buying. The "Adulteration of Food Act" has, however, been very efficacious in putting an end to this form of adulteration. Lard, again, was at one time very largely used as an ingredient of adulteration, the dishonest vendor obtaining the extra profit resulting from the difference in value between the lard and butter.

99. **IMITATIONS OF BUTTER.**—The adulteration of butter having been pretty effectually stopped, the ingenious manufacturers of the article made from refuse fat and suet now sell it, presenting all the appearance of genuine butter, under the name of "butterine." Large quantities of this imitation of butter are now sold, at prices varying from sixpence-halfpenny per pound to a shilling per pound, it being put up in the tub form and wearing the look of proper butter. The article was at one time comparatively unknown in England, but recently large factories have been established for its manufacture; the result being that the butchers throughout the kingdom have now no difficulty in disposing of their refuse fat and stale suet.

100. **BUTTER AS FOOD.**—Butter, like milk, is a universal article of food, its component parts assisting the human economy when used in proper moderation, and few morning or evening meals would be considered complete without it in the great majority of households throughout the United Kingdom.

There is always a current sale for butter; and, as an established article familiar to everyone, there is no danger of its not continuing to be held in public favour as one of the most necessary items of the daily food of the people.

101. **MARKETS.**—Every town and every village throughout the

kingdom of any size is a market for butter, where the price rules according to the quality of the article, and its position as respects nearness or distance from the seat of supply. The chief cities and large manufacturing towns all present a never-satiated market, London especially, and there is never any trouble in disposing of large quantities of really good butter.

102. **IMPORTATIONS.**—The demand for butter is so great and universal that large quantities are annually sent over to this country from France, Holland, and America. Dutch butter, as an article of medium quality, has long held a fair place in public estimation, the imports reaching to a very considerable amount. We do not get so much butter from America as cheese, yet their system of making butter in the States is very perfect and complete, and some valuable hints are to be obtained from it by the English butter-maker.

103. **AMERICAN FACTORIES FOR BUTTER MANUFACTURE.** —The American system of associated dairies has been described by Mr. X. A. Willard, A.M., of Herkimer, New York, in the pages of the *Journal of the Royal Agricultural Society :*—

The plan was first originated in 1851 by Jesse Williams, who planned the first cheese factory, and the system is found by American dairymen to produce as much extra profit as would suffice to pay for the entire cost of management under the individual system, the result being a constant improvement in dairy management. At first cheese-making only was designed by Mr. Williams, but his success induced the butter dairymen of Orange County, New York, so to modify his system, as to render it applicable to the production of butter. For nearly ninety years the whole farming population of Orange County have directed their chief attention to butter-making and the production of fresh milk for the New York market, and the associated system has caused the methods for obtaining the cream, and the produce itself, to attain the highest degree of excellence, and long prices are paid for it.

What is termed " fancy butter " will fetch a dollar a-pound, and can only be produced from very superior pastures. The old pastures in the district referred to embrace the June or blue grass *(Poa pratensis),* the fowl meadow grass *(Poa serotina),* meadow fescue *(Testuca pratensis),* red top *(Agrostis vulgaris),* the wire grass *(Poa compressa),* the sweet-scented vernal and vanilla grass *(Dactylis glomerata),* clover, and other forage plants.

The June grass *(Poa pratensis)* is regarded as very valuable : it throws out a dense mass of leaves, is highly relished by cattle, and produces milk from which a superior quality of butter is made. The wire grass *(Poa compressa)* is deemed one of the most nutritive of the grasses, is very hardy, eagerly sought after by cattle, and is one of the best grasses for fattening. Cows feeding upon it yield milk of the richest quality, from which the nicest butter is made. It flourishes well upon gravelly knolls and in shaded places, and its stem is green after the seed has ripened. It is found growing in all the States of the Union.

The meadow fescue is common in old grass lands where the sod is thick, and grasses of different varieties are mingled together. It starts up early in the spring, is relished by stock, and furnishes good early feed. The milk-farmers hold it in high estimation as a reliable grass, tenacious of life, and not running out like timothy *(Phleum pratense)*, or clover. The white clover *(Trifolium repens)* springs up spontaneously in the old pastures, and is highly esteemed, as giving quality and flavour to butter.

The sweet-scented vernal grass grows best upon the moist soil of the old meadows. It starts very early, and gives off an agreeable odour.

We have named the grasses quoted by Mr. Willard; but probably soil and climate may modify their character in other places.

No particular breed of cattle is in special favour in the United States, and amongst those on an American dairy-farm are found Jersey or Alderney cows, Short-horns, Ayrshires, Devons, as well as those having a dash of Holstein blood in them, obtained by crossing thoroughbreds upon the common cows of the country. The herds on a farm average about 25 cows; some carry 40 to 60, but in the majority of cases the herds are small, ranging from 15 to 30 cows.

The cost of erecting a good factory, and supplying it with machinery, is about 4,000 dollars (£800).

The milk, as soon as it comes from the cow, is strained and put into long tin pails, which are set in cold spring water, care being taken that no portion of the milk in the pails be higher than the flowing water which surrounds it, in pools constructed for the purpose. These pails are 8 inches in diameter, and from 17 to 20 inches long.

The milk is stirred occasionally, to prevent the cream from rising. It is important that the animal heat should be removed from the milk as soon as possible, at least in an hour's time after it has been drawn from the cow.

The old method was to cool the milk in the large carrying-cans, but it has been found that it keeps sweet longer by dividing it into small quantities and cooling it in pails as above described. The milk stands in pails surrounded by fresh spring water until ready to be carted to the trains; it is then put into carrying-cans holding from 40 to 50 gallons. The cans are completely filled, and the covers, which fit closely, are adjusted so that there shall be no space intervening between them and the milk.

One of the principal features of the American system are these pools of water, sunk below the level of the floor, into which the pails of milk are placed, which are filled to within four inches of the top. The best temperature of the water for the purpose is considered about 56° Fahr. The pools, it is considered, should not be kept at so low a temperature as 48°, nor much, if any, above 57°.

It is claimed that more cream, and that of better quality for butter-making, may be obtained by setting the milk on the above plan than it will yield in shallower pans, or when exposed to uneven temperatures. (Pails 20 to 22 inches in length and 8 inches in diameter.)

Another feature deemed of great importance is to expose as little of the surface of the milk to the air as possible, in order that the top of the cream may not get dry, which has a tendency to fleck the butter and injure the flavour. The milk of one day is left in the pools until next morning, which gives 24 hours for the morning's mess, and 12 hours for the evening's mess to cream. The pails are then taken out of the pools and the cream dipped off.

In the fall and spring of the year the cream, as it is dipped, goes immediately to the churn and is churned sweet; in summer the cream is dipped into the pails and returned to the pool, and kept there till it acquires a slightly acid taste, when it is ready for the churn.

The cream having been removed, the skimmed milk in the pails is now turned into the cheese-vat to be made into "skim-cheese."

In some factories where an extra fancy product of butter and skimmed cheese is desired, none of the milk is set longer than 24 hours, and at these factories it is not desired to take all the cream from the milk, but only the best part, and employ the remainder to give extra quality to the "skim-cheese."

The churning is done by horse-power, the churn most commonly used being simply a large circular platform, or wooden wheel, built about an upright shaft,

G

the lower end of which turns in a socket. The wheel sets upon an incline, so that
the horse, by walking constantly on one side, keeps it in motion. At the upper
end of the shaft gearing is arranged, so as to give motion to the churn. The old-
fashioned barrel dash-churn is generally liked in America. Four dash-churns
are sometimes placed side by side, so as all to be worked by power at the same
time. From 60 to 70 quarts of cream are put into each churn, and each mess of
cream then receives from 12 to 16 quarts of water, for the purpose of diluting it
and bringing it to a temperature of about 60°. Cold spring water is used in
warm weather, and warm water in cold weather.

Some prefer diluting the cream with water, and passing it through a sieve
before putting it in the churns, in order that the particles of cream may be all of
uniform size, since, if the butter does not come evenly, but is mixed with small
particles of cream, it will soon deteriorate, and will not make a prime or fancy
article, as it is termed. This point is considered of great importance by the best
butter-makers, and it is claimed that the method of setting the milk in deep pails,
by which all thin cream is obtained, rather than the thick, leathery masses
skimmed from milk set in pans, renders it more evenly churned, and thus secures
a better product.

In warm weather, ice is sometimes broken up and put in the churn to reduce
the temperature of the cream; but it is deemed better to churn without ice, if
the cream does not rise above 64° F. in the process of churning, as butter made
with ice is more sensitive to heat. It is, however, a less evil to use ice than to
have the butter come from the churn white and soft. In churning, the dashes
are so arranged as to go downwards within a quarter of an inch of the bottom of
the churn, and to rise above the cream in their upward stroke.

The temperature of the cream while being churned should be kept below 65°,
for if, at the close of the churning, the butter-milk should be at that temperature,
or above it, the flavour and colour of the butter will be injured. In cold
weather, the temperature of the cream, when ready for churning, is a little
higher than in warm weather, about 62° being considered the right point.

104. **SALTING.**—Salting butter is often confusedly managed in
England without any distinct reference to the exact quantity of
salt used. In America, when the butter has been removed from
the churn, and care taken not to touch it more than is absolutely
necessary with the hands, salt is added, and worked through the
butter with the butter-worker at the rate of 18 oz. of salt to 22 lbs.
of butter. Great care is taken that the salt be pure, and of those
brands that are known to be good. For butter that is designed to
be kept over for the winter markets, a little more salt is sometimes
used, often as high as an ounce of salt to the pound of butter.
Not unfrequently a teaspoonful of pulverised saltpetre and a tablespoonful of white sugar are added at the last working for 22 lbs. of
butter.

In the matter of salt, however, the factories adapt the quantity
to suit the taste of their customers, or for the different markets.
Of late years light salted butter sells best in America, and the rate of
salting varies from one-half to three-fourths of an ounce of salt to
the pound of butter. The butter, after having been salted and
worked, is allowed to stand until evening, and is then worked a

second time and packed. In hot weather, as soon as the butter is salted and worked over, it is taken to the pools and immersed in water, where it remains until evening, when it is taken out, worked over, and packed. For this purpose a separate pool is provided, which is used only for butter. It is called the "butter pool," and fresh spring water constantly flows in and out of it, as in the pools for setting the milk.

105. **PLAN OF WORKING THE BUTTER.**—In working the butter, considerable skill and experience are required, that its grain shall not be injured. The butter must have a peculiar firmness and fineness of texture, and a wax-like appearance when fractured, which an improper handling in expelling the butter-milk and working will destroy. Care is taken, therefore, not to overwork it, nor subject it to a grinding manipulation like tempering mortar, as this spoils the grain, and renders the butter of a greasy or salvelike texture.

The butter is worked with butter-workers. The one in most common use consists of an inclined slab standing upon legs, and with bevelled sides about 3 inches high. The slab is 4 feet long by 2 wide at the upper end, and tapering down 4 inches at the lower end, where there is a cross-piece, with a slot for the reception of the end of the lever. There is also an opening at this end for the escape of the butter-milk, with a pail below. The lever is made either with four or eight sides, and the end fits loosely in the slot, so as to be worked in any direction. It is quite simple, but does good execution, and is much liked in the butter factories.

106. **ADVANTAGES OF BUTTER FACTORIES.**—The advantages of butter-making on the associated-dairy system over that in private families is very great. In the first place, by the association system a uniform product of superior character is secured. Every appliance that science, or skill, or close attention is able to obtain, is brought to bear upon the manufacture, and prime quality necessarily follows as a result.

If you could assume that in a neighbourhood of a hundred families, each family had the skill and convenience of the factory, and that each would give the subject the same close attention, doubtless there would be no difference as to the quality of product; but such a state of things rarely exists.

Again, the factories are able to obtain a larger price, because it costs the dealer no more time to purchase the hundred dairies combined than it would to purchase an individual dairy, and the uniformity and reliability of the product does not entail the losses that are constantly occurring in different small lots by reason of inferior quality. The factories, too, relieve the farmer and his family from a great deal of drudgery, and unless the work can be done by members of the family who cannot be employed profitably at other labours, it is a matter of economy to have the butter and cheese made at the factory, since what would take a hundred hands scattered over the country to do, is performed in the same time by three or four when the milk is worked up together in one place. The only serious complaint against the factory system is in hauling the milk. This has been obviated, in many instances, by establishing a route of milk-teams, where milk is delivered for the season by the payment of a small sum.

107. SKIM-CHEESE.—The manufacture of skim-cheese is a part of the American butter-factory system, the cream being dipped from the milk while it is sweet, and the latter then goes into the milk-vats for making " skim-cheese."

In making a " fancy " product it is found advisable that the delivery of milk be kept within moderate bounds, say from 300 to 400 cows. The factory milk-vats are all essentially alike in form and size. They hold from 500 to 600 gallons.

There is a great variety of heating apparatus, boilers, steamers, tanks for hot water, and what is termed " self-heaters," that is, with fire-box attached to and immediately below the milk-vat. This kind of heater is very popular at the butter factories, as it consumes very little fuel, is easily managed, and does as good work as the best.

The ordinary heater is constructed separately from the vat, and consists of wrought-iron pipes screwed together in such a manner as to form a fire-chamber, and present a large amount of heated surface.

Where a boiler and engine are used, power is afforded for driving the churns, and in this respect this system must prove most convenient. Still, as the expense is considerably more than for the self-heater, both in the first cost and for fuel, many prefer the latter.

108. RESULTS OBTAINED AT THE BUTTER FACTORIES— LABOUR, &c.—The average product from the milk during the season at the butter factories is a pound of butter and two pounds of skim-cheese from 14 quarts of milk. There is a variation in the quality of milk at different seasons of the year; and in the fall, when the cows are giving a smaller quantity, it is, of course, richer in cream, and better results are obtained from the same quantity than early

in the season. This will be seen from the following examples of a single day's work, taken at random from the book of one of the factories :—

On May 18th, from 3,512 quarts of milk, wine measure, there was produced 213 lbs. of butter and 560 lbs. of skim-cheese. On May 26th, from 3,300 quarts of milk, 210 lbs. of butter and 550 lbs. of cheese. On September 12th, from 3,180 quarts of milk, 200 lbs. of butter and 546 lbs. of cheese. On October 14th, from 2,027 quarts of milk, 120 lbs. of butter and 407 lbs. of cheese.

In the working of any system practical men always desire statistics of results. The following is a statement of receipts and expenditure at one of the small butter factories, where a portion of the milk was sold :—

The quantity of milk received from April 10th to December 1st was 627,174 quarts, of which 27,308 were sold at a little above 7 cents ($3\frac{1}{2}d.$) per quart, leaving 509,866 quarts to be made up into butter and cheese.

The product was as follows :—31,630 lbs. of butter, 81,778 lbs. of skim-cheese, 15,908 lbs. whole-milk cheese, 2,261 quarts of cream, sold at $19\frac{1}{16}$ cents ($9\frac{8}{10}d.$) per quart, and 1,561 quarts of skim-milk.

The net cash receipts, after deducting transportation and commissions, were as follows :—

	Dollars.
For pure milk sold	1,926,22
For skim milk sold	24,02
For butter sold	13,344,21
For skim cheese sold	11,659,08
For whole-milk cheese	1,065,44
For 2,261 quarts cream	443,33
Hogs fed on whey	446,24
Butter-milk and sundries	207,49
Making total of	29,116,03

(Equal to £5,823 4s. 1½d.)

The expense account was as follows :—

	Dollars.
For labour	1,476,40
For fuel	79,96
For cheese boxes	653,17
For 20 sacks of salt	89,25
For rennets, bandages, &c.	483,55
For carting cheese and butter to station	273,10
Paid for hogs	179,90
Total	3,235,33

(Equal to £647 1s. 4d.)

This gives an aggregate net receipt of 25,880,70 dollars.

From these statements it appears that the butter averaged 42¼ cents (say 1s. 9d.) per lb., the skim-cheese 14¼ cents (about 7d.), and the whole-milk cheese 18 cents per lb., while the average amount received on the whole quantity of milk was 4$\frac{1}{10}$ cents (2d.) per quart. The whole expenses of the factory were a little over ½ cent per quart.

For working this factory there were employed, besides the superintendent, three hands, viz., two men and one woman. The labour account for conducting this factory, it will be seen, is a little over two mills $\frac{7}{10}$ per quart.

As will be seen from the foregoing, everything is conducted upon the closest calculation and most complete system, to ensure a definite result. Not only is great economy practised in the cost of producing the butter and cheese, but the prices realized are very high when a first-rate article is turned out. They are, indeed, fancy prices, but such as the best families in New York are, or were, at all events, accustomed to give.

It will also be seen in some essential particulars the American plan is different from the English; as, for example, in exposing so little of the surface of the milk to the atmosphere, when the custom here is generally to expose as much as possible, with a view to the more complete rising of the cream to the top.

In England the *quantity* of dairy produce that can be turned out is the chief point aimed at, in many instances too little attention generally being paid to the *quality*, though sometimes, where the farmer's wife herself superintends the operations of the dairy, a first-rate article is turned out, by reason of the care and attention that is bestowed upon it. And, although of late years American competition has been loudly complained of, it is not at all improbable that, if some of the finest dairy produce in the shape of Gloucester, Stilton, and other cheeses, &c., were sent over to New York, a market could be found for it even there at remunerative prices, though at first sight it might appear somewhat like "sending coals to Newcastle."

109. **COST OF PRODUCTION AND PROFIT.**—From what we have already written, it will be seen how much the cost of production varies under different conditions, and profits, of course, are affected in the same ratio, but the average proportion of milk, cream, and butter to each other, is 1 gallon of cream to 9 of milk, and 3 pounds of butter to 1 gallon of cream, or 1 pound of butter to 2½ or 3 gallons of milk as it comes from the cow. The result of the latter

will of course depend upon the richness of the milk, a much larger quantity of butter being obtainable from Alderney cows in proportion to the amount of milk yielded, than from cows which give a large quantity of milk, necessarily poorer, according to measurement, of cream, or the butter-making properties.

The following is an estimate of the cost and annual produce of a cow in a dairy district in Scotland, where the cows were highly fed. The price of the butter would seem low to many, but in agricultural districts, where it is sold off in large quantities to dealers who buy the whole produce of a dairy, often not more than 11½d. per lb. is obtained, though the same butter, perhaps, sold in small quantities to consumers, would readily fetch 16d., or even more.

Expense, from May 1st to October 1st.

	£	s.	d.
2 acres of grass, at 45s.	4	10	0
Clover and tares	1	0	0
Draft in summer	0	6	5½

From October 1st to May 1st.

	£	s.	d.
14 tons 4 cwt. of turnips, at 7s. 6d.	5	6	6
5 bushels of linseed, at 7s.	1	15	0
Draft in winter	1	2	1
Interest on £14 at 5 per cent.	0	14	0
Carriage of milk, and tolls	0	15	0
Attendance, fuel, &c.	0	10	0
Total	£15	19	0½

Produce per cow = 680 gallons of milk.

	£	s.	d.
227 lbs. of butter, at 10½d.	9	18	7½
600 gallons skimmed milk, at 4½d.	11	5	0
50 gallons of butter-milk, at 2¾d.	0	9	4½
Calf at a week old	0	15	0
Total value of produce	£22	8	0
Deduct expense of food, &c.	15	19	0½
Net profit per cow	£6	8	11½

In addition to the above, the value of the manure must be taken into consideration, which is much greater in the case of highly-fed cows than in that of poor ones.

CHEESE-STOOL.

CHAPTER IX.

CHEESE.

Utensils used in making Cheese—Cheshire Cheese—Double Gloucester Cheese —Stilton Cheese—Dunlop Cheese—Cream Cheese and Skim-Milk Cheese— Colouring Cheese—Rennet—Salting, Drying, &c.—Mites and Flies—Mouldy Cheese—Importation of Foreign Cheese—Cost of Production and Profits.

110. **UTENSILS USED IN MAKING CHEESE.**—The utensils used in cheese-making vary but little throughout the different counties of the kingdom, though the processes of making cheese are very diversified, and are included in the milk-pail, cheese-tub, sieve, cheese-vat, and circular board, skimming-dish and bowl, and cheese-press. These last are of several forms; sometimes a block of stone, or a box full of stones, let up and down by rope and pulley, or by windlass, upon three or four cheeses, one above another, under it. As it is of importance to regulate the pressure, the lever cheese-press is considered best.

The Milk-Pail is generally supposed to hold six gallons, maple being thought the best wood for the purpose. The cheese-tub is of a capacity sufficient to hold the milk from which the cheese is intended to be made. Cheese-vats are of various sizes, being usually turned out of solid elm. In Gloucestershire, where five cheeses go to the cwt., in the formation of the familiar "Double Gloucester," the vats are 15½ inches in diameter by 4½ inches deep; "Single Gloucester," of which it takes eight cheeses to make a cwt., are of the same diameter, but only 2¾ inches deep; the only real difference between the two being the size of the cheese, and the difference of quality arising from the longer period during which the thicker cheese must be kept in order to ripen.

III. **THE METHODS OF MAKING CHEESE IN ENGLAND** have not varied in any essential particulars for a great length of time, except in the more fanciful kinds, which, of late years, have been introduced into the market.

The method of making Cheshire cheese is thus described in *Holland's Survey :—* "Take about a pint of cream, when two-meal cheeses are made, from the night's milk of twenty cows. In order to make cheese of the best quality, and in the greatest abundance, it is, however, admitted that the cream should remain in the milk ; for whether the cream that is once separated from it can by any means be again so intimately united with it as not to undergo a decomposition in the after process, admits of some doubt. The more common practice is, however, to set the evening's milk apart till the following morning, when the cream is skimmed off, and three or four gallons of the milk are poured into a brass pan, which is immediately placed in the furnace of hot water, and made scalding hot ; then half of the milk thus heated is poured upon the night's milk, and the other half is mixed with the cream, which is thus liquefied, so as, when put into the cheese-tub, to form one uniform fluid. This is done by the dairywoman while the other servants are milking the cows, and the morning's milk being then immediately added to that of the evening, the whole mass is at once set together for cheese.

"The rennet and colouring being then put into the tub, the whole is well stirred together, a wooden cover is put over the tub, and over that is thrown a linen cloth. The usual time of 'coming,' or curdling, is one hour and a half, during which time it is frequently to be examined. If the cream rises to the surface before the coming takes place, as it often does, the whole must be stirred together so as to mix again the milk and the cream ; and this as often as it rises, until the coagulation commences. If the dairywoman supposes the milk to have been accidentally put together cooler than she intended, or that its coolness is the cause of its not coming, hot water or hot milk may be poured into it, or hot water in a brass pan may be partially immerged in it. This must, however, be done before it is at all coagulated, for the forming of the curd must not be tampered with. If it has been set together too hot, the opposite means, under the same precautions, may be resorted to ; but the more general practice is to suffer the process to proceed, hot as it is, until the first quantity of whey is taken off, a part of which, being set to cool, is then returned into the tub to cool the curd. If too little appears to have been used, it renders the curd exceedingly tender, and therefore an additional quantity may be put in ; but this must be done before the coagulation takes place, for, if added afterwards, it will be of little effect, as it cannot be used without disturbing the curd, which can then only acquire a proper degree of toughness by having some heated whey poured over it.

"Within an hour and a half, as already mentioned, if all goes on well, the coagulation will be formed—a point which is determined by gently pressing the surface of the milk with the back of the hand ; but in this test experience is the only guide, for the firmness of the curd, if the milk be set hot together, will be much greater than that from milk which has been set cold together. If the curd be firm, the usual practice is to take a common case-knife and make incisions across it to the full depth of the blade, at the distance of about one inch, and again crosswise in the same manner, the incisions intersecting each other at right angles. The cheese-maker and two assistants then proceed to break the curd, by repeatedly putting their hands down into the tub and breaking every part of it as small as possible, this part of the business being continued until the whole is uniformly broken small ; it generally takes up about forty minutes, and the curd is then left, covered over with a cloth, for half an hour, to subside.

"The bottom of the tub is now set rather atilt, the curd is collected to the upper side of it, and a board is introduced of a semi-circular form to fit loosely one-half of the tub's bottom. This board is placed on the curd, and a bolt weight upon it, to press out the whey, which, draining to the lower side of the tilted tub, is ladled out into brass pans. Such parts of the curd as are pressed from under the board are cut off with a knife, placed under the weighted board, and again pressed ; the operation being repeated again and again, until the whey is entirely drawn from the curd. The whole mass of curd is then turned upside down, and put on the other side of the tub to be pressed as before. The board and weight being removed, the curd is afterwards cut into pieces of about eight or nine inches square, piled upon each other, and pressed both by the weight and hand ; these several operations being repeatedly performed as long as any whey appears to remain in it.

"The next thing is to cut the curd into three nearly equal portions, one of which is put into a brass pan, and is there by two women broken extremely fine, a large handful of salt being added and well mixed with it. That portion of curd being sufficiently broken is put into a cheese-vat, which is placed to receive it on a cheese-ladder over the cheese-tub, the vat being furnished with a coarse cheese-cloth. The second and third portions of the curd are heated in the same manner, and emptied into the vat, except that into the middle portion eight, nine, or ten times the quantity of salt is usually put. By some dairy-women, however, each portion is salted alike, and with no more than three large handfuls to each. The breaking takes up more or less time, as the cheese was set together hotter or colder ; half an hour is perhaps the longest time.

"The curd, when put into the cheese-vat in its broken state, is heaped above the vat in a conical form ; to prevent it from crumbling down, the four corners of the cheese-cloth are turned up over it, and the women, placing their hands against the conical part, gently, but forcibly, press it together, constantly shifting their hands when any portion of the curd is starting from the mass, and folding down the cloth upon it. As soon as the curd adheres together so as to admit of it, a small square board, with a corner of the cloth under it, is put on the top with a 60-lb. weight, or a lever is pressed upon it. Several iron skewers are at the same time stuck in the cone, as well as through holes in the side of the vat, from which they are occasionally drawn out and fixed in other spots, until not a drop of whey is discharged. The weight and skewers are then removed, and the corners of the cloth are either held up by a woman or by a wooden hoop, while the curd is broken as small as possible, half-way to the bottom of the vat ; and the same operation of pressing and skewering is repeated. The women then take up the four corners of the cloth while the vat is drawn away and rinsed in warm whey ; a clean cloth is then put over the upper part of the curd, and it is returned inverted into the vat. It is then broken half-way through in the same manner as before, which several occupations occupy from three to four hours.

"When no more whey can be extracted by these means from the cheese, it is again turned in the vat, and rinsed as before in warm whey. The cloth now made use of is finer and larger than the former, and is so laid that on one side it shall be level with the edge of the vat, and on the other wrap over the whole surface of the cheese, the edges being put within the vat, thus perfectly enclosing the entire mass. In this stage of the business the cheese is still higher than the edge of the vat ; and to preserve it in due form, recourse is had to a binder, about three inches broad, either as a hoop, or as a cheese-fillet, which is a strong, broad, coarse sort of tape, which is put round the cheese, on the outside of the cloth, and the lower edge of the binder pressed down within the vat, so low as that the upper edge of it may be level with the surface. The cheese is then carried to the press, and a smooth strong board being placed over it, the press is gently let down upon it, the usual power of which is about 14 or 15 cwt. In most dairies, however, there are two presses, and in many three or four of different weights; the cheese being by some put first under the heaviest, and by others under the lightest.

"As soon as the cheese is put into the press, it is immediately well skewered; the skewers being of strong wire, 18 or 20 inches long, sharp at the points and broad at the other end; the vat and binder having holes, seldom more than an inch asunder, to receive them. As the press always stands near the wall, only one side of the cheese can be skewered at the same time, and it must therefore be turned half-way round, whenever that is necessary; but this occasions no inconvenience, as the skewers must be frequently shifted, and many more holes are made than skewers to fill them. In half an hour from the time the cheese is first put into the press, it is taken out again, and turned, in the vat, with another clean cloth, after which it is returned to the vat; but is by some persons previously put naked into warm whey, where it stands an hour or more for the purpose of hardening its coat. At 6 o'clock in the evening the cheese is again turned in the vat into another clean cloth, and some dairywomen prick its upper surface all over an inch or two deep, with a view of preventing blisters. At 6 o'clock on the following morning it is again turned in the vat, with a clean cloth as before, and the skewers are laid aside; it is also turned two or three times more, both morning and evening, at the last of which finer cloths are used than those at first, in order that as little impression as possible may be made on its coat.

"After the cheese has remained about forty-eight hours under the press, it is taken out, a fine cloth being merely used as a lining to the vat, without covering the upper part of the cheese, which is then placed nearly mid-deep in a salting-tub, its upper surface being covered all over with salt. It stands there generally about three days, is turned daily, and at each turning well salted, the cloth being changed twice in the time. It is then taken out of the vat, in lieu of which a wooden girth or hoop is made use of, equal in breadth to the thickness nearly of the cheese, and in this it is placed on the salting-bench, where it stands about eight days, being well salted all over, and turned each day. The cheese is then washed in lukewarm water, and after being wiped, is placed on the drying-bench, where it remains about seven days; it is then again washed and dried as before, and, after it has stood about two hours, it is smeared all over with about two ounces of sweet whey-butter, and then placed in the warmest part of the cheese-room.

"While it remains there it is, during the first seven days, rubbed every day all over, and generally smeared with sweet butter; after which it should for some time be turned daily, and rubbed three times a week in summer, and twice in winter. The labour is performed almost universally by women, and that in large dairies, where the cheeses are sometimes, upon an average, of 140 lbs. each, and the whole of this process refers to cheeses of large size and to extensive dairy operations."

112. **DOUBLE GLOUCESTER CHEESE.**—Welsh rare-bits, or "Welsh rabbits" as they are more commonly called by the majority of persons who are not very particular as to the derivation of terms, were perhaps more generally in request a good many years back than they are at the present time, and toasted cheese was more commonly eaten, and was a more general dish than it now is; and the various qualities of different cheeses used to be studied with a view to their toasting properties.

Double Gloucester was for a long time celebrated for this purpose, the mild-ness of its flavour, combined with its great richness and that adhesive nature which permits it to be cut in slices without crumbling, causing it to be peculiarly suitable, until "Single Gloucester," or "toasting cheese," was made, being of a size well adapted for slices for toasting, the weight of a cheese seldom exceeding 12 lbs., while that of Double Gloucester is generally about 22 lbs., and the mode of making it is just the same as that followed in making Double Gloucester. Occasionally, however, it is not made so rich, and there is less salt put in it, while it is pressed only four days instead of five. Substantially it is the same as Double Gloucester, the method of making which has been thus described by Mr. Hayward, who used to have an extensive dairy at Frocester Court:—" When the curd is sufficiently firm for breaking, it is gently and slowly cut crosswise to the bottom of the tub, at about an inch apart, with a three-bladed knife of fourteen inches long. When it has stood five or ten minutes, to allow it to sink a little, and the whey to come out as clear as possible, some of the whey is dipped out of it with a bowl, and the curd is again cut. This must also be at first done slowly, and with strokes at a considerable distance from each other; for, if performed hurriedly, a great sediment of curd will be found in the whey-leads; it should, however, be gradually quickened, and the strokes taken nearer and nearer every time, one hand with skimming-dish keeping the whole in motion and turning up the lumps suspended in the whey, while the other cuts them as small as possible. This process may occupy a quarter of an hour.

" The curd is now allowed to settle during a quarter of an hour, when the whey is taken from it and poured through a very fine hair sieve placed over the whey-leads; the dairymaid then cutting the curd into lumps, from which most of the remaining whey escapes. The curd is then pressed down with the hand into vats, which are covered with large cheese-cloths of fine canvas and placed in the press for half an hour, after which they are taken out and the curd put into a mill of Mr. Hayward's construction, which tears it into small crumbs, and saves the laborious part of squeezing and rubbing it with the hands, while it also retains that portion of the oily matter which would be otherwise lost to the cheese, and thus occasions a great improvement in the making.

" In this pulverised state it is customary with most dairymaids to scald the curd with hot whey; but Mrs. Hayward considers the cheese richer when not scalded, for this washes out a part of the fat; she therefore merely presses it closely together with the hand when filling the vat. The whey should, however, be completely extracted, and the curd filled into the vat as compactly as possible, being rounded up in the middle, but only just so much as that it can be pressed down to a level. A cheese-cloth is then spread over the vat, and a little hot water is thrown over the cloth, as tending to harden the outsides of the cheese and prevent it from cracking. The curd is now turned out of the vat into the cloth, and the inside of the vat being washed in whey, the inverted curd, with the cloth around it, is again returned to it; the cloth is then folded over, and the vat put into the press, where it remains about two hours, after which it is taken out and dry cloths applied, which should be repeated in the course of the day; it is then replaced in the press until the cheese is salted, which is generally done within twenty-four hours after it is made.

" The salting is performed by rubbing the entire of the cheese with finely-powdered salt, for if the curd be salted before being put into the vat, its particles do not intimately unite, and although it may become a good cheese, it is loose and crumbly, and never becomes a smooth, close, solid mass like that which is salted after it has been made ; but this is never done until the skin is closed, for if there be any crack in it at that time it will not afterwards close. The cheese is, after this, returned to the vat and put under the press, in which more cheeses than one are placed together, care being always taken to put the newest lowest in the press, and the oldest uppermost. The salting is repeated three times, the cloths being removed after the second in order to efface their marks, and twenty-four hours are allowed to intervene between each; thus the cheese is within five days taken from the press to the cheese-room; though in damp weather it should remain somewhat longer. There it is turned every day for a month, when it is ready for cleaning, which is done by scraping with a common knife, the dairymaid sitting on the floor and taking the cheese in her lap to perform the operation. When it has been cleared from all scurf, it is rubbed all over with a woollen cloth dipped in paint made of Indian red, or Spanish brown, and small beer; and as soon as the state of the paint will permit, the edge of the cheese, and about an inch on each side, are rubbed hard with a cloth every week. The quantity of salt is generally about 3½ lbs. per cwt., and one pound of annatto is sufficient for half a ton of cheese."

113. **STILTON CHEESE.**—Takes its name from the place where it was first made, near Melton, in Leicestershire, though they are now commonly made in several other counties besides Leicester, as those of Cambridge, Huntingdon, and Nottingham. They are of small size and richer in quality than most cheeses, having more cream put into them than most others have. As Stilton cheeses are often given as presents, a large trade is done in them, particularly at Christmas time, or rather before.

It is made by putting the night's cream to the milk of the following morning, or, if the cheese is desired to be very rich, a still greater proportion of cream. The rennet is then added, but no colouring matter is used; and when the curd has come, unlike the method pursued in making most other cheeses, it is taken out without being broken and put whole into a drainer, where it is squeezed down hard until the whey is entirely pressed out. When dry it is put, with a clean cloth, into a chessel, and placed beneath the press, the outside being first well salted. When sufficiently firm to be removed, it is put upon a dry board, and tightly bound round with a cloth, which must be changed daily, in order to avoid cracks in the skin, until it is found to have a coat formed, when there is no occasion for its further use, and nothing more need be done to the cheese than to brush it occasionally, and frequently turn it upside down upon the shelf or stand where it may be placed.

114. **DUNLOP CHEESE.**—Has acquired a good reputation in the market, and is now very generally made in the counties of Lanark, Renfrew, Ayr, and Galloway. They are put up in moderate sizes, varying from 28 to 56 lbs. The process of making has been described by Mr. Aiton.

"When so many cows are kept on one farm as that a cheese of any tolerable size may be made every time they are milked, the milk is passed, immediately as it comes from them, through a sieve into the vat, and, when the whole is collected, it is formed into a curd by the mixture of the rennet. Where, however, the cows are not so numerous as to yield milk sufficient to form a cheese at each meal, the milk of another meal is stored about six or eight inches deep in coolers, and placed in the milk-house. The cream is then skimmed from the milk in the coolers, and, without being heated, is put into the curd-vat along with the milk just drawn from the cows, and the cold milk, from which the cream has been taken, is heated, so as to raise the temperature to about blood-heat. This, indeed, is a matter of great importance; and though in summer 90° may be sufficient, yet, upon the average of winter weather, 95° will be generally found requisite. If coagulated much warmer, the curd becomes too adhesive, much of the butyraceous matter is lost in the whey, and the cheese will be found dry, tough, and tasteless; but if too cold, the curd, which is then soft, does not part readily with the serum, and the cheese is so wanting in firmness that it is difficult to be kept together; indeed, even when the utmost pains are taken to extract the whey, and give solidity to the cheese, holes—which, in dairy language, are termed 'eyes,' 'whey-drops,' and 'springs'—frequently break out, and always render them either rancid or insipid.

"About a tablespoonful of the liquid rennet is generally thought sufficient for 100 quarts of milk, and the curd is usually formed by it within twelve or fifteen minutes; though in some dairies—of course in consequence of the difference in strength in the rennet—it does not come for from three-quarters of an hour to an hour, though double the quantity of rennet is used. The curd is then broken with the skimming-dish or with the hand, and the whey ought to be taken off as speedily as possible, though without pressing, as the least violence has been found to make it come off white, and thus weaken the quality of the cheese. (The best method of separating the whey from the curd, as recommended in the *Trans. of the Highland Society,* is, in the first instance, to lift the edge of the cheese-tub, and let the whey run off slowly from it into a vessel placed underneath. The tub is then let down to stand a little, after which it is turned one-fourth round, and another collection emptied off. Thus, by turning the tub a fourth round every time, it is found to part from the curd more pure and quickly.)

"When quite freed from the whey, and the curd has acquired a little consistence, it is then cut with the cheese-knife—gently at first, and more minutely as it hardens; after which it is put into the drainer (which is a square vessel with small holes in the bottom, and a cover to fit inside), on which the lid is placed, with a cloth thrown over it, and a slight pressure, say from three to four stones weight, according to the quantity of curd—being laid on, it is allowed to stand from fifteen to twenty minutes or half-an-hour. It is then cut into pieces of two inches square, the whey is again discharged, and the weight, being doubled, is replaced. This process of cutting is smaller every half-hour, and, increasing the weight until the pressure is upwards of 100 lbs., is continued for three or four hours. It is then cut very small, and minutely salted; half-an-ounce of salt, or, at the most, thirteen ounces to twenty-four pounds English, being sufficient.

"A clean cheese-cloth, rinsed in warm water and wrung out, being placed in the chessel, the curd is then put into it, and a half-hundredweight laid on it for an hour. It is then put under a press of two hundredweight, where it remains during an hour and a half; after which it is taken out, and, a fresh cloth being placed in the chessel, the cheese is turned upside down, and laid, with increased weight, under the press during the whole night. Next morning, and during the three or four days which it must remain in the press, it is daily turned repeatedly, dry cloths being each time used, and the weight is gradually increased until the pressure amounts to at least a ton.

"When ultimately taken from the press, the cheeses are generally kept during a week or ten days in the farmer's kitchen, where they are turned three or four times every day, and rubbed with a dry cloth. They are then removed to the store-room, which should be in a cool exposure, between damp and dry, without the sun being allowed to shine upon them, or yet a great current of air admitted—this gradual mode of ripening being found essential to prevent the fermentation and heaving of the cheese, as well as the cracking of the rind; but attention must be paid to rub them with a dry cloth and turn them daily for a month or two, and twice every week afterwards.

115. **CREAM CHEESE AND SKIM-MILK CHEESE.**—These two cheeses are the exact opposites to each other with respect to the relative component parts of which they consist, the one being all cream, and the other all skimmed milk.

Cream Cheeses are indeed little more than portions of thick, sweet cream which have been dried by being placed in a miniature cheese-vat of about an inch and a half in depth, with small holes at the bottom, through which any residue of milk can drain. It is covered with rushes, or the stalks of Indian corn, so placed as to allow of the cheese being turned without handling it, and it is never pressed at all, except very gently by the hand between cloths. It is then placed in a somewhat warm situation to ripen and sweat. If the frost touches it, it becomes spoiled, and loses its taste and flavour. On the other hand, if kept too hot it acquires a rank taste, and extreme heat must, on this account, be guarded against.

Skim-milk Cheese, as its name implies, is made from milk from which all the cream has been removed. There are various qualities of skim-milk cheese, the worst being very indigestible; and this depends chiefly upon the time the milk has been allowed to stand. If it has stood so long as to be deprived altogether of the butyraceous matter, it is very poor stuff.

It used to be made in very large quantities in Suffolk (being known by the name of " Suffolk bang "), where at one time it had such an unenviable reputation that it was asserted it used to be chopped up with a hatchet instead of being cut with a knife; or, if a man wanted a bit of stick to fasten up a gate with, and could not find a piece of wood handy, he would cut a wedge off his luncheon cheese for the purpose and make use of it. In old times, when the farm labourers lived partially or wholly in the house with the farmer, the quality of the cheese used often to become a bone of contention, being at times too hard to bite; so that it used humorously to be said the labourers in that part of the country, having to " bolt " their cheese in blocks, by a long course of practice had acquired *square throats.*

To make skim-milk cheese of fair quality, the milk, if possible, should not be allowed to become sour, and as soon as it has been skimmed it should not be made warmer than animal heat, or about 90°, for if put together too hot it will turn out very tough; and as the curd coagulates much quicker than that of whole milk from which the cream has not been removed, there is no necessity for causing it to have the same degree of heat. This is the principal item in the difference of management, except that it is more difficult to break the curd, and the cheese wants less pressing. It will be also much sooner ready for use than the whole-milk cheese, not requiring to stand so long.

116. COLOURING CHEESE.—Cheese is commonly coloured with Spanish annatto, which is generally used by rubbing a piece of it in a bowl with some warm milk, which is afterwards allowed to stand

for a short time in order to draw off the sediment. A piece of annatto weighing rather more than a quarter of an ounce is sufficient to colour a cheese of the weight of 60 lbs. Marigolds boiled in milk are used by some persons to colour cheese, and this is a favourite method with many; while others employ carrots, also boiled in milk, and strained, which imparts a rich-looking colour to the cheese, but gives it rather a definite taste, on which account their use is often avoided.

117. RENNET.—Rennet is prepared by different methods in different districts. In some districts the contents of the stomach of a calf are preserved with salt, and used, but this method is somewhat repugnant to many. In some of the midland counties the cleaned stomach of a calf is salted, pickled and dried, and when at least a year old it is well soaked in salt and water, half-a-pint of which is enough for fifty gallons of milk. In Cheshire, the skins are cleaned out and packed in salt till the following year. A month or so before they are wanted, three or four inches are steeped during the night in half-a-pint of salt and lukewarm water, for use in the morning, and put with fifty or sixty gallons of milk. Cheese can be made from the curd formed by the coagulation of the milk when it turns sour, but this is not so effective, and causes it to be hard and of indifferent flavour, and does not nearly answer so well as the gastric juice that is found in the "maws" or stomachs of calves that have been fed entirely on milk. The more usual method is to use the skins of the stomach-bag alone, which are rolled up with salt, and hung up in a warm place to dry, after which they are put aside for a long time before they are used. If the skin be good, a small piece not larger than a nut, soaked for twelve hours in a cupful of water, is enough for twenty gallons of milk. The quality of the cheese depends very much upon the proper application of the rennet. If the maws or "vells" are too new (twelve months being considered the earliest date at which they are fit for use, after first being selected for use) they cause the cheese to heave, or swell, which makes it full of "eyes" or holes. If too much, again, is used, or if it be unusually strong, it will also cause the cheese to heave by inducing fermentation. The vells of pigs and lambs have been found amongst those sent from Ireland, but these do not answer the purpose so well, the Irish calves' vells being considered the best by many for this purpose. The somewhat nauseous idea which attaches itself to

This necessary operation—i.e., the application of rennet—has been disguised

by the addition of spices and sweet herbs. Here is an old receipt from the West of England :—

" When the rennet-bag is fit for the purpose, let two quarts of soft water be mixed with salt, wherein should be put almost every sort of spice and aromatic herb that can be procured, and it must boil gently till the liquor is reduced to three pints ; it should then be strained clear from the spices and poured in a tepid state upon the mass, and a lemon may be sliced into it, when it may remain a day or two, after which it should be strained again and put in a bottle, where, if well corked, it will keep good for twelve months or more, and give the cheese a pleasing flavour."

In Marshall's " Southern Counties," the method recommended is as follows:— "Take the maw of a newly-killed calf and clean it of its contents : salt the bag, and put it into an earthen jar for three or four days, till it form a pickle ; then take it from the jar and hang it up to dry, after which it is to be replaced in the jar, the covering of which should be pierced with a few small holes to admit the air, and let it remain there for about twelve months.

" When wanted for use, a handful each of the leaves of sweet-briar, dog-rose, and bramble, with three or four handfuls of salt, are to be boiled together in a gallon of water for a quarter of an hour, when the liquor is to be strained off and allowed to cool. The maw is then to be put into the liquid, together with a lemon stuck round with cloves, and the longer it remains in it the stronger and the better will be the rennet : half-a-pint, or less, of the liquor is sufficient to turn fifty gallons of milk."

The method of preparing rennet in Cheshire, described by Holland, is thus :— "When the maw comes from the butcher, it is always found to contain a chyley, or curd-like matter, which is frequently salted for present use, but when this chyley matter is taken out, and the skin cleared from slime and every apparent impurity by wiping, or a gentle washing, the skin is then filled nearly full of salt, and, placing a layer of salt upon the bottom of a mug, the skin is placed flat upon it. The mug is large enough to hold three skins in a course, each of which should be covered with salt ; and when a sufficient number of skins are thus placed in the mug, it should be filled up with salt and put, with a dish or slate over it, into a cool place till the approach of the cheese-making season in the following year. The skins are then all taken out, laid for the brine to drain from them ; and, being spread upon a table, they are powdered on each side with fine salt, and are rolled smooth with a paste roller, which presses in the salt. After that, a thin splint of wood is stuck across each of them to keep them extended while they are hung up to dry.

" In making the rennet, a part of the dried maw skin is, in the evening previous to its being used, put into half-a-pint of lukewarm water, to which is added as much salt as will lie on a shilling. In the morning this infusion (the skin being first taken out) is put into the tub of milk ; but so great is the difference in the quality of these skins, that it is difficult to ascertain what quantity will be necessary for the intended purpose. A piece the size of half-a-crown, cut from the bottom of a good skin, will commonly be sufficient for a cheese of 60 lbs. weight, though ten square inches of skin are often found too little. It is customary, however, to cut two pieces from each skin, one from the lower, the other from the upper part ; but the bottom end is the strongest.

" *An improved mode is :*—To take all the maw-skins provided for the whole season, pickled and dried as before ; put them into an open vessel, and for each skin pour in three pints of spring water ; let them stand twenty-four hours, then take out the skins and put them into other vessels ; add for each one pint of spring water, and let them stand twenty-four hours as before. On taking the skins out the second time, gently stroke them down with the hand into the infusion : they are then done with. Mix these two infusions together, pass the liquor through a fine linen sieve, and add to the whole a quantity of salt, rather more than is sufficient to saturate the water ; that is, until a portion of salt remains undissolved at the

H

bottom of the vessel. The next day, and also the summer through, the scum, as it rises, is to be cleared off, and fresh salt should be added. Somewhat less than half-a-pint of this preparation will generally be sufficient for 60 lbs. of cheese; but, when for use, the whole should be well stirred up."

There are occasionally (though, fortunately, it is of rare occurrence), in the course of some preparations of our daily food, details which necessarily must be attended to, but which wear a somewhat repugnant aspect. The modern fine lady who perhaps enjoys a roast fowl for her dinner, would not relish her meal so much if she had to "draw" the bird before it was cooked; and these expedients of adding spices and sweet herbs to rennet, by which our kindly great-grandmothers invested a somewhat unsavoury piece of business with sweeter surroundings, we ought to be grateful for; but the result of these applications is quite unimportant as far as the making of cheese is concerned.

118. SALTING.—Some apply salt to cheese in twelve hours, but this is considered too soon, and it is thought best to do so after the cheese has been twenty-four hours in the press, when it is ready for receiving it, for as a general rule the salt should not be applied until the skin of the cheese is firm and free from openings, as these never close so completely after salting, whatever amount of pressure may be applied.

The salting is done by the hand, the salt being rubbed over the entire surface of the cheese for as long as it will absorb it, after which it is wrapped up again in a dry cloth and put under the press. Twenty-four hours afterwards it is salted again as before, this time being put into the vat without a cloth and pressed, in order to produce a smooth and even surface.

A final rubbing of salt is given once again after the same interval, and the cheese being pressed as before is ready for removal to the drying-room.

119. DRYING, STORING, &c.—It is important to have a special dry-room, or loft, set aside for cheese, into which the cheeses as they are removed from the press should be taken, and laid either upon shelves, racks, or on the floor, where they are easily accessible, so as to be well wiped with dry cloths and turned every twelve hours for three days. After the first three days they need only to be wiped and turned every twenty-four hours, and in a month after leaving the press they are ready for being scraped. When cheese is intended for the London market it is generally painted at this time, the paint used being Indian red or Spanish brown, or a

mixture of both with table beer, which is rubbed on with a woollen cloth.

120. **MITES AND FLIES.**—When the cheeses are being turned in the drying-room, they should be closely examined while being regularly turned, and cleared from mites. In warm weather the flies are apt to attack cracks or soft parts of the cheeses, and when this takes place the best plan is to scoop out very thoroughly the affected part so as to leave nothing suspicious behind, and fill it up again with the soft part of another cheese kept for the purpose, and cover carefully with cloths. Attention to these details will raise the standard and character of the produce.

121. **MOULDY CHEESE.**—Skimmed milk often becomes blue moulded, which is generally much relished and considered a great improvement to the taste. This is occasioned sometimes by cracks in the cheese, where the mould-plant vegetates and spreads through the whole mass.

Mouldiness is sometimes artificially produced by pouring port wine into holes bored in the cheese, and by exposing it to a damp, close atmosphere. Again, if by accident a little sour milk has been used in making the cheese, mouldiness invariably ensues. When the best quality of cheese has become mouldy, it is considered by many a great delicacy; it is highly stomachic, and a corrective after fruit has been eaten.

122. **IMPORTATION OF FOREIGN CHEESE.**—A good deal of cheese is sent to us from Holland, large quantities of the familiar Dutch cheese being especially sold in London, while Switzerland sends certain fancy kinds, as well as France, the latter importations consisting mostly of soft kinds, which are looked upon as delicacies, and consumed to a large extent in the best London dining-houses and hotels; but the great bulk of foreign cheese comes to us from America, and is very various in quality, it being manufactured in the States upon a large routine system, in a similar manner to that described as done with butter, which we regret our space will not allow us to give a description of, but of which a tolerable estimate may be gathered from that we have already referred to.

123. **COST OF PRODUCTION AND PROFITS.**—An account of the cost of production and profits upon cheese-making must necessarily be only approximate, as, with care and attention to details in feeding and management, the profit in one case will be double that in another. We give, however, what has been considered the money profit of a cow in Gloucestershire, upon which calculations

used to be based in large cheese-making dairy farms, prices being fixed upon a low scale, and the money value of the cow at £16:—

COST.			£	s.	d.
Grass and hay		...	9	0	0
Attendance, milking, and cheese-making...		...	1	10	0
Deteriorated annual value (cow kept five years)...		...	1	4	0
Insurance, 4d. per £ on £16...		...	0	5	4
Interest on capital, £16 (5 per cent.)		...	0	16	0
			£12	15	4

PRODUCE.			£	s.	d.
500 gallons of milk, made into cheese at 6d. per lb. ...		...	12	10	0
20 lbs. cream-butter, at 11d.		...	0	18	4
30 lbs. whey-butter, at 9d.		...	1	2	6
Whey for feeding pigs, say...		...	1	5	0
Calf sold at a week old		...	0	10	0
	Total produce	... ...	16	5	10
	Deduct	...	12	15	4
	Net Profit	...	£3	10	6

We give these figures merely as an approximate method of reckoning, and as it takes a gallon of milk to make a pound of cheese, and milk at the lowest will fetch 8d. per gallon, it would not be worth while to make it into cheese and sell it for 6d.! But as, on account of difficulties connected with situation or otherwise, it may be necessary to make the produce of a dairy into cheese, the items of expenditure and profit will very much depend upon the skilful management of the person interested. As we have shown before, cheese-making is the least profitable of all the systems of dairy-farming.

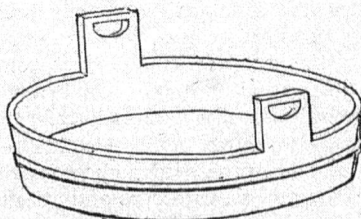

BUTTER-TUB.

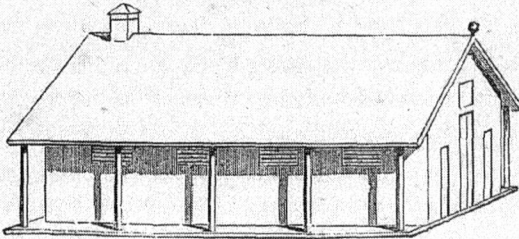

DUTCH COW-HOUSE.

CHAPTER X.

BREEDING.

Improvement of Breeds—Calves—Birth of the Calf—Hard Udder—Recipe for Sore Teats—Artificially Feeding the Calf—Hay Tea and Linseed Jelly—Skimmed Milk as Food for Calves—Importance of Regular and ample Feeding to Young Stock—Shed for Calves—One Cow will Suckle Five Calves—Method of Rearing Calves in Ireland—Weaning the Calf—A Good and Cheap Food—Castrating.

124. **IMPROVEMENT OF BREEDS.**—A judicious breeder has it in his own power very much to develop those points in his stock that he wishes to see them possessed of, if he takes the necessary pains to do so; and he must, of course, in the first place, make up his mind as to the points he intends to aim at in the breeding, to suit the ultimate purpose he has in view.

If he wishes only to rear cows for dairy purposes, he must sell off the bull calves dropped by those breeds of cows which make the best milkers, supposing they have been crossed by a bull of a similar breed, such as the Ayrshire and Alderney, whose characteristics we have before described, as they will not grow up into favourable stock for the grazier, or butcher, or answer his own purpose to fatten ultimately.

If, on the other hand, the improvement of stock is aimed at, the object first to be considered is to obtain animals which will yield the largest return in the shortest time from the consumption of the food they have given to them, and experience proves that in cattle-breeding the qualities of the calf are mostly influenced by the male parent, and thus a uniform quality of stock is to be obtained.

If flesh-forming animals are wanted, there is nothing to excel the shorthorn, but high-pedigreed shorthorn cows are not good milkers, and they are not usually desirable for breeding in the general way, except by those who make breeding a business, and want a fine race; a less refined-bred cow being better for ordinary purposes. Crossed by a shorthorn bull of the best breed, some good calves are thus to be procured, and in choosing the cows, in the first place, animals of large frame and vigorous constitution should always be selected.

125. **CALVES.**—If the calf is not intended to remain with its mother, as soon as it is dropped it should be removed to the calf-

house, and placed in a well-littered crib, and immediately rubbed all over with straw wisps to remove the mucus with which it is covered. The dam always performs this office herself in the most effectual manner in a state of nature, and it is best, we consider, to leave the calf a short time with its mother, as we shall afterwards describe; but the plan we are now referring to is carried out in those cases where several calves are to be reared simultaneously on the milk of one cow, for this reason—when the calf is removed at birth, without allowing the dam to see it or lick it, she frets less than when it is allowed to remain with her for a short time and then removed; and she gives her milk more freely when milked by the hand from the first.

At first it is necessary to feed the calf with its own dam's milk, which nature endows at the time with a peculiar quality that acts as a purge to the calf, and clears its bowels of the meconium they are charged with at birth; but this first milk, or " beestings " as it is commonly called, must on no account be given to older calves, to which it would be hurtful.

We will, however, describe from the beginning what we consider a better method of rearing calves economically and yet effectually.

126. BIRTH OF THE CALF.—When the cow's term of gestation is nearly complete, she should be kept in a quiet place near the house, and it will be of great assistance to her in calving if her bowels are opened by a dose of medicine, which will cause her to have an easier time.

As her time of parturition draws near, it will be evidenced by symptoms of uneasiness and moaning, accompanied by a dropping of the belly, the springing of her udder, and a discharge from the bearing. In the event of severe weather she should be housed, and a good bed made for her. Cold water should be kept out of her reach, and, in ordinary cases, Youatt recommends that a pint of sound warmed ale be given to her in an equal quantity of gruel; and warm gruel should be frequently administered, or, at all events, placed within the animal's reach; and in ordinary cases, where there is only some little delay, and nothing serious apprehended from a wrong position of the fœtus, to the first pint of ale should be added a quarter-of-an-ounce of the ergot of rye (spurred rye), finely powdered, and the same quantity of ergot with half-a-pint of ale should be repeated every hour until the pains are reproduced in their former and natural strength, or the labour terminated.

After calving, a warm mash should be put before her, and warm water, or water from which the chill has been taken off, two or three hours after which it will be advisable to administer an aperient draught, consisting of a pound of Epsom salts and two drachms of powdered ginger. If the placenta, or after-birth, is not soon discharged from the body, the aperient draught should be given together with the ergot of rye and ale.

Cows eat the after-birth, or " cleansing," which, it is supposed, is designed by nature to act as a medicine ; but it is very often taken away and put aside, as being too disgusting to be allowed to remain.

In cases of difficulty, unless a very experienced man is on the farm, it will be safest to send for a veterinary surgeon ; but all going on well, the cow should be left quietly with her calf, the licking and cleaning of which will amuse her.

Whatever is done with the calf ultimately, it should at least be left with the cow for three or four days. As Youatt justly says : "It is a cruel thing to separate the mother from the young so soon; the cow will pine, and will be deprived of that medicine which nature designed for her, in that moisture which hangs about the calf, and even in the placenta itself, and the calf will lose that gentle friction and motion which helps to give it the immediate use of all its limbs." The calf also derives a benefit from the first milk of the cow, which possesses an aperient quality.

127. **HARD UDDER; RECIPE FOR SORE TEATS.**— In the case of young cows the udder is often hard, and the calf should then be allowed to suck for a fortnight, and if the first calf, left to suck until old enough to wean. After a short interval the cow should be milked by hand first, so that the calf gets the last milk, which is the richest, and the udder is softened in the attempts made by the calf to obtain it.

It sometimes happens that the teats of the cow become sore, and she manifests a disinclination for the calf to suck her, in which case they should be fomented three or four times a-day with warm water, after which she should be very carefully and gently milked. The teats should then be dressed with an ointment, which can be readily made, composed of an ounce of yellow wax, and three of lard, melted together. When these begin to get cool, well rub in a quarter-of-an-ounce of sugar of lead, and a drachm of alum finely powdered. Should the cow not readily commence to lick off the slimy matter with which the calf when first born is covered, if a handful of common salt is sprinkled over it she will generally perform this duty at once. Some farmers make a practice of giving the calf lukewarm gruel instead of the " beestings," or first milk of the cow, which is a wrong practice, as the calf loses the benefit of the aperient quality we have before alluded to, which assists in removing the glutinous fæces which have accumulated in its intestines.

Youatt recommends that when the calf has been cleaned and has begun to suck, the navel-string should be examined, and if it continues to bleed a ligature should be passed round it closer, but, if it can be avoided, not quite close to the belly. Possibly the spot at which the division of the cord took place may be more than usually sore ; a pledget of tow, well wetted with Friar's balsam, should then be placed over it, confined with a bandage, and changed every morning and night, but the caustic applications which are so frequently resorted to should be avoided.

128. ARTIFICIALLY FEEDING THE CALF; HAY TEA AND LINSEED JELLY.—A principal object of the present work being to show how these farm operations can be economically conducted, so as to create a larger margin of profit, we will mention that calves can be reared with but little expense as store-calves, if the necessary trouble is taken with them. It is sound policy to allow them to remain with the cow for a week, so as to give them a fair start, as it were, and during this first week with its dam, four quarts of milk per day, at two meals, is sufficient. After this it can feed very well upon skimmed milk, so that the farmer can get his usual quantity of butter and rear his calf into the bargain. Many substitutes for milk have been given, with more or less success ; hay tea and linseed jelly being the most resorted to. Linseed jelly is made by putting one quart of seed to six of water, and allowing it to boil for ten minutes. Hay tea is made by pouring boiling water over fine sweet hay, and enclosing the vessel—generally a large earthen pan with a lid—and in a couple of hours a strong liquid is produced. It should be given of the warmth of the natural milk of the cow, and, if given without, linseed should be mixed with three-parts of skimmed milk, and be afterwards reduced to one-fourth.

129. SKIMMED MILK AS FOOD FOR CALVES.—No better way, however, can be found of disposing of the skimmed milk than feeding calves with it. It wants a careful man to feed them, the milk not being allowed to get the least sour, or the calf will scour, and be thrown back very much in its progress. It should be boiled, and given the natural heat of the milk as it comes from the cow, and either thickened with linseed or oatmeal. The calf can easily be taught to drink from the pail. At first it will not know what to make of it, but if the man wets his fingers with the milk and places them to the calf's mouth inside the pail, the little animal will soon get an inkling of the business in hand.

As with the cow, regularity of feeding is imperatively demanded if the calves are expected to thrive, and they should be fed at least three times a-day. If seldom fed, the calves will drink fast and become "paunchy" when their food is given only at morning and evening, which it is the practice of some farmers to do, allowing them upon these occasions to thoroughly satiate their appetites, which fills the stomach and impedes digestion, which is obviated by more frequent feeding. After the fourth week the calf will begin to eat a little sweet green hay, and, a couple of weeks later on, sliced roots, meal, or finely-crushed cake. Nothing, however, will beat the skimmed milk, thickened with meal or linseed. Some very successful calf rearers use a mixture of linseed and ground wheat to thicken with, in the proportion of two bushels of linseed to one of wheat.

130. **IMPORTANCE OF REGULAR AND AMPLE FEEDING TO YOUNG STOCK.**—There is one very important point which should always be steadily kept in mind by those who aim at rearing stock successfully, and that is, from the very first birth of the animals to push them steadily forward to condition by careful, regular, and sufficient feeding—not extravagant feeding, but a sufficiently liberal amount of food should be given to all young stock to ensure their steady progress. Any check given to this progress may retard it for months afterwards. Those who have half-starved their animals for a length of time cannot profitably atone for their neglect or bad management by sudden and lavish attempts to push them on to fatness. If the proper treatment and feeding of animals is neglected it is sure to result in loss and disappointment to the owner, and the only way to make them pay is to see their growth and improvement continued without cessation from the earliest period of their existence, till they are either killed or sold in the market. Expensive food in large quantities has often had to be given to neglected animals, which, although they ate it greedily, like Pharaoh's lean kine, for a long period seemed to do them little good, its cost being also very considerable.

It will be seen that if the method of rearing calves to the best advantage is followed, *i.e.*, feeding them upon skimmed milk, there will be lost to the pigs a very important item of their food, and it must be left to the farmer to say which he attaches the most importance to. One thing, the pigs can be fed without the skimmed milk, which the calves ought not to be deprived of; and if stock is wanted, the skimmed milk will put them in a fair way of being acquired at a trifling cost.

131. **SHED FOR CALVES.**—Calves reared in this manner should be placed in a warm shed. No matter if they come at the inclement season of Christmas or January; this season perhaps is rather better for them than not, as they will be growing into strength against the time when the spring comes round, and will so enjoy the benefit of every day of fine weather after they are turned out till the autumn, when they will have changed into large strong animals. Almost any shed or outhouse can soon be converted into a capital house for

calves, so that it is warm and dry, with the floor sloping in order that the urine may flow towards a cesspool. The outhouse should be separated by divisions, which can be made by hurdles, one end fastened to the wall and the other pegged down to the ground. The calves should be fastened with a halter-rope to a ring, through which it is allowed to play, these rings being driven into a piece of quartering nailed to the wall at a height of two and a half feet from the floor. One hay-rack of the common old-fashioned semi-circular kind will serve two calves, and should be placed at a height of three-and-a-half feet from the floor.

The young calf when about two months old will begin to nibble a little grass, and as soon as the weather is fit it should be let out for a couple of hours or so on sunny days, supposing it to be born about the commencement of the year, but should not be allowed to eat too much. In three months or so it will have acquired a taste for grass, when its feed of milk, &c., can be gradually discontinued. The calf by another month or two will be a strong healthy animal in most cases, and will give no further trouble.

One cow will suckle five calves.—When farmers wish to bring up calves by allowing them to suck the cow, it has been considered the best plan to let them suckle two pairs in succession and one afterwards, making five calves in all. This is managed in the following manner, described by Youatt:—

"A strange calf is purchased which is put along with her own to the same cow, both being put to suck, one at each side, exactly at the same time, and leaving them there for fifteen or twenty minutes, by which time the milk will be drawn away. The cow at first shows great dislike to the stranger, but in a few days receives it very quietly. They are thus kept in the house, and as they advance in age they eat porridge, hay, sliced potatoes, or any food that is usually given to them, and in about three months they are finally turned out to grass; after which a couple of strangers are purchased, and the same plan pursued with them during three months longer. At the expiration of that period—supposing the cow to have calved in the month of January or the early part of February—the first week in August will have arrived, and this set being then ready for weaning, a single calf is put into the feeding-pen, and fattened for the butcher, by which means the cow will have suckled five calves."

132. METHOD OF REARING CALVES IN IRELAND.—A very similar method to the one described of rearing calves economically has been related by Mr. Hooper as followed in Ireland.

The reader will perceive there are certain points of resemblance in the systems carried on at distances far apart from each other, though possibly some portions of them may appear a little whimsical, as, for example, beating up an egg in the calves' food.

The calf being dropped, it should be borne in mind that from that moment until its arrival at maturity it must be kept progressing,

improving, growing; in no other way will it pay. The question is, how is this best to be done?

133. **WEANING THE CALF.**—We now come to the weaning of the calf. I find my calves do best in a clover stubble, that is, on the stubble of a barley or oat field that is laid out with clover and grass seeds; and I do not find that such light stock do any harm to the young clovers, provided they are not kept on them longer than the first of January. On whatever kind of pasture they are weaned, however, the milk should not be taken from them too suddenly; they should get a meal of milk in the middle of the day for a week or ten days after being turned out to grass, and the quantity of this should be diminished by degrees. When the nights begin to get cold, generally some time in the month of October, they should be housed at night; but, in my opinion, should be allowed out by day the whole winter, and, therefore, should not be kept in too close or warm a house by night. I have kept them in close houses, and only let them out for a few hours on fine days; and I have kept them in altogether, day and night, the whole winter, some tied and some in loose boxes; but I find them thrive and grow best on the system I have adopted for the last three or four years, and that is, to tie them up at night in a shed open to the south, and let them out the whole day in all weather, excepting, of course, when the ground is covered with snow, and they could get nothing to eat. If they could be put into a yard at night, with sheds around it, perhaps it would be better still, as some have an objection to tying up young calves; but if it has its disadvantages, it has also advantages, one of which is that they are much quieter when tied up the next winter, at which age I fatten off my young stock; whereas I find those I buy take some time to get accustomed to the chain and trough, and lose time accordingly. As to the feeding of calves, the first winter I find they do better, with the outrun I have spoken of, on hay alone, than on hay and straw, and turnips. Some may think this unreasonable, but I can only say that I have wintered calves for eight years *with* turnips, and for four years *without* them, and I have no intention of altering my present system.

134. **A GOOD AND CHEAP FOOD.**—Mr. Burke, whom I quoted before, who rears calves with so little milk, winters them on pulped mangolds, mixed with straw, chaff, oil-cake, and crushed corn, and he says, reckoning 5 cwt. of straw chaff at 5s., 10 cwt. of pulped mangolds at 5s., 1 cwt. of oilcake at 10s., and 4 cwt. of mixed crushed corn at 30s., he has one ton of food for 50s. equal to the best hay.

But as our hay is seldom worth even as much as this in our own yards, I do not think we should gain much here by the adoption of his system, which, however, is certainly a great improvement on the ordinary one of whole or sliced turnips and hay or straw. I make no difference in the wintering of calves intended for beef and those intended for the dairy. No matter what a calf is intended for, it should be well fed the first winter, or it will receive a check from which it will never recover. To proceed with my own system. I give my yearlings the best grass I have all the summer, and fatten them off the following winter; selling them when 24, 25, and 26 months old. I give them oilcake to the amount of 30s. a head (beginning with 1 lb. a day, and increasing gradually to 3 lbs.), and hay or straw and turnips *ad libitum;* and the best fetch from £18 to £20 a head, and the smaller ones from £15 to £16, that is at the present and recent high price of beef. My cattle get three feeds of sliced turnips in the day; the first between 5 and 6, the next at 11 (immediately before which they get their oilcake), and the third and last at 5, or later as the days get longer. At each of the meals, if any animals have finished before the others, or show any desire for more, more is given till they are satisfied. The cleaning out of the stall, and currying of the cattle, keep them disturbed a good part of the time between the first two meals; but after the second meal they are left to rest till the third, and after that they are left undisturbed for the night. The racks placed above the turnips are filled with hay or straw as often as may be required. At whatever age cattle are put into the stall, they should have some turnips given them in the grass for a week or two before they are put in, to accustom them to a change of food. The heifers that appear best adapted for the dairy I sell as springers in October and November, when two and a half years old. In my opinion this is young enough for even the largest and best heifers to calve, and smaller and weaker ones should not calve till they are nearly or fully three years old, according to their size and strength."

135. CASTRATING.—The bull calves should be castrated when about three weeks old, as at this age it is done with less risk and, to all appearance, with less pain to the animals than when done when they are older. At one time it was the practice to spay female calves, but of late years it has been abandoned as hazardous and useless.

SCOTCH HIGHLAND BREED.

CHAPTER XI.

CATTLE.—PASTORAL FARMING.

Qualifications required for Pastoral Farming—Capital—Suitable Localities—Varieties of Cattle—The Short-horned Durham Ox—Fifeshire and Lowland Scotch—Herefords—The Galloway, Aberdeen, Angus, and small Scotch breeds—The Ayrshire—The Long-horns—The Alderney—Welsh Cattle—Advice on Buying—Accommodation for Cattle—Feeding—Straw as Food for Cattle—Summer Feeding—Winter Feeding.

136. QUALIFICATIONS REQUIRED FOR PASTORAL FARMING.—There are certain branches of the subject with which we are now dealing that necessarily become somewhat divided under different headings, where a numerous range of subjects have to be treated upon in a comparatively small space, in which the various breeds and different methods of feeding stock profitably have to be alluded to, and of course many of the remarks which will follow relative to the economical feeding of the ox, and that of the cow under the head of dairy management, will indifferently apply to either.

With respect to the different breeds of cattle from which choice has to be made, this should depend upon their adaptation to the soil on which it is designed to place them, and the food that is intended to be given, either in the form of grazing, soiling, or stall-feeding. Care should be taken in the selection of stock, if intended for grazing, that an animal is not placed on inferior pasture to the kind upon which he has been reared or accustomed. Where there are fertile meadows with rich bottom herbage, the large heavy breeds of cattle, or any other kind, are sure to do well; and whether it is sound meadow land, rough and indifferent pasture, or marsh, makes all the difference, and the beasts should be selected with the view of getting the most fitted and suitable kind for the land that is farmed. On poor land, where the herbage is scanty, some of the Highland breeds will answer very well, where the larger species, such as the Hereford and short-horned Yorkshire, would almost starve. Some of the smaller kinds of Scotch cattle are contented with the scantiest herbage ; and although these are often small in size, where a good cross is wanted to be obtained, the progeny raised from one of these small cows and a short-horned bull often turn out remarkably well, and of good size and handsome proportions. The Kyloes, or West Highland cattle, do very well upon the coarsest pasturage, and are a handsome breed to boot, and its choice quality of meat causes the breed to be a favourite one with butchers. The cows give but little milk, and soon become dry, but the milk is of first-class quality. It should ever be borne in mind that to fatten lean cattle ought never to be attempted, unless they are in good order as store cattle to start with, and for this purpose moderate-sized beasts, weighing from 40 to 60 stones of 14 pounds each, are the most saleable. The smaller kinds of Scotch and Welsh cattle are to be bought cheap at the different fairs which the drovers attend, and some very useful animals are often to be picked up which turn out remarkably well eventually.

137. CAPITAL.—The capital to be employed in pastoral farming will of course depend upon the scale upon which operations are conducted. Considerably less will be required than in arable culture, and its amount will have to be regulated by the number of beasts that the aim is to rear and fatten, and the cost of their attendance, which is comparatively small.

As we have shown, where cows are kept calves can be reared at a comparatively small expense, and useful stock for fattening is to be bought out of the droves of Irish, Welsh, and other cattle that are to be met with at the various fairs. The cost of calves and young stock thus purchased, according to age and varying conditions, may be put down as ranging from £5 to £10. The cost of any given number of stock may thus be roughly estimated, to which must be added a year's keeping of the animals, the cost of wages for their attendance, miscellaneous petty expenses, and a year's rent.

138. SUITABLE LOCALITIES.—Cattle rearing is now successfully carried out upon an immensely extended range of soils to what used formerly to be the case, many farms now yielding an annual lot of fat bullocks which many years back used to produce none, chiefly attributable to the great increase in turnip cultivation and the free use of auxiliary feeding substances ; but it is very desirable to have on every farm where cattle are bred a sheltered paddock near the homestead, into which young calves can be turned with as little trouble and loss of time as possible.

Comfortable quarters and generous diet are the first essentials of rearing young stock, and where these are the best and easiest to be obtained is the chief consideration to be entertained in making choice of a locality.

139. VARIETIES OF CATTLE.—As we have previously pointed out, the nature of the pasture should be taken into account where grazing is largely carried on, for some of the hardier kinds of cattle do very well upon coarse or scanty herbage, that would be quite unsuitable for the larger and more highly-bred animals.

In breeding, one has the opportunity of rearing whatever kind of cattle may be desired, and where dairying operations are carried on, one great advantage there is, that although the cows, themselves good milkers, are not of the breed whose progeny make good oxen for the butcher where the bull is also the same, yet it is the case that these, when coupled with a shorthorn bull, even in cases of cows of diminutive size, will produce a cross which rivals in weight of carcase some of the largest kinds.

140. **SHORT-HORNED DURHAM OX.**—The breed commonly known as "shorthorn" is considered the best and most profitable we have in England, arriving early at maturity and supplying meat to the butcher of the best quality. Under the local name of "Teeswaters" this breed got a firm hold on public estimation in the county of Durham towards the close of the last century, and the stock has been steadily kept improving. Their origin is somewhat uncertain, some maintaining that they sprang from Dutch extraction and were imported into Hull, while others contend they can be traced to the Western Highlands, and have mixed, or Kyloe blood in them, but this has been denied.

Mr. Henry H. Dixon, in a Prize Essay upon the " Rise and Progress of Shorthorns," which appeared in the Journal of the Royal Agricultural Society, in which the fullest particulars are given relative to various celebrated animals and their pedigrees, after giving a long list of well-known names, says :—" The germ of this wonderful array must have been considered an 'improved' county breed as far back as 1787. Hutchinson, of Sockburn, had then a cow good enough to be modelled for the cathedral vane, and had also beaten Robert Colling in a bull class. . . . . 'Hubback' (319) has always been considered the great regenerator of shorthorns, but he did not do Charles Colling so much good as 'Foljambe,' who was from a 'Hubback' cow, and he was parted with at the end of two seasons. The aim of the brothers Colling was to reduce the size and improve the general symmetry of their beasts," &c. &c.

To the bull in question ("Hubback") most breeders used to be desirous of tracing their stock. Mr. George Coates, an eminent breeder who first collected the pedigrees of short-horned cattle, gives the following particulars which he obtained respecting this celebrated beast from the person from whom he received them, which are embodied in the following letter :—

" I remember the cow, which my father bred, that was the dam of 'Hubback'; there was an idea that she had mixed, or Kyloe blood in her. Much has lately been said that she was descended from a Kyloe, but I have no reason to believe, nor do I believe, that she had any mixture of Kyloe blood in her.

"(Signed)   JOHN HUNTER.

"Hurworth, near Darlington.
    "July 6th, 1822."

And in Mr. Coates' Herd Book is registered the following :—"'Hubback' (319), yellow, red, and white, calved in 1777, bred by Mr. John Hunter, of Hurworth; got by Mr. George Snowdon's bull (612), his dam (bred by Mr. Hunter) by a bull of Mr. Bankes, of Hurworth. g.d., bought of Mr. Stephenson, of Ketton."

Mr. Dixon says that it is calculated by experienced Smithfield salesmen that rather more than two-thirds of the average number of beasts (331,164) which came to the London market so long ago as 1863-64 were either pure shorthorns, or shorthorn crosses. In reference to this increase, an old English breeder writes :—" When I began there was no pure-bred shorthorn bull within seventeen miles of me, whereas now there is one in every parish."

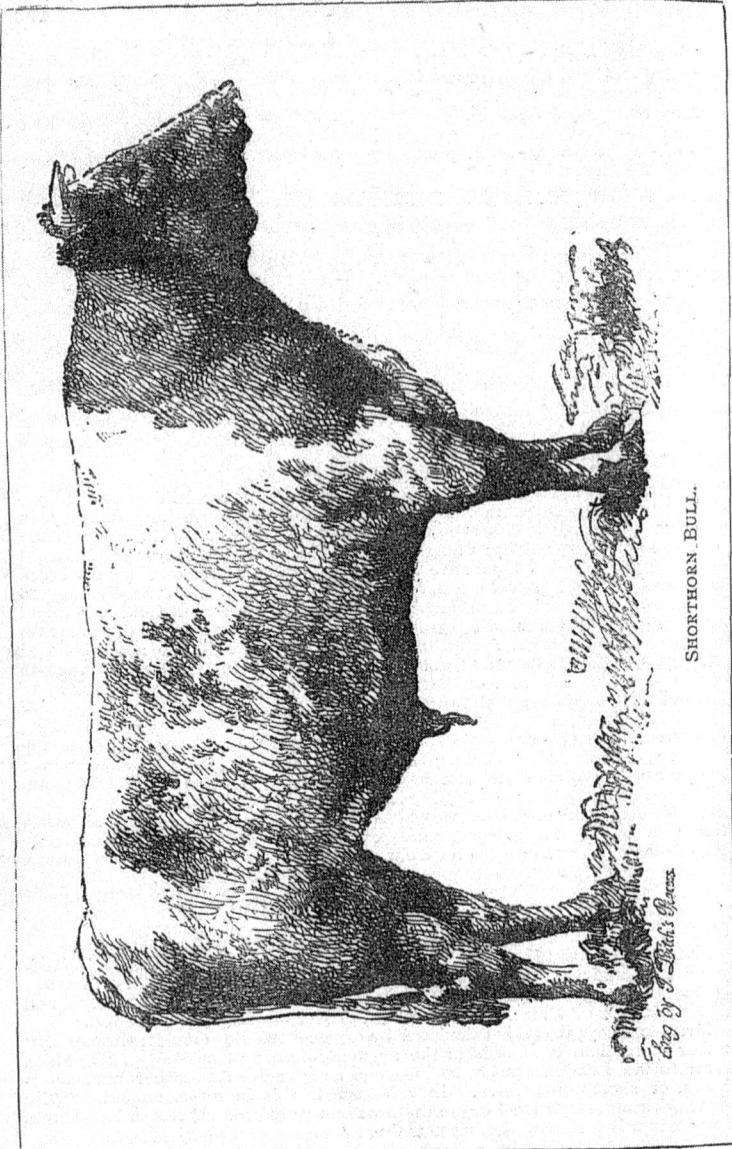

SHORTHORN BULL.

Eng by J. Bird's Sons

The merits, indeed, of the shorthorn breed are indisputable, for steers of from four to five years old, weighing 140 stone of 14 pounds, and sometimes as high as 150 stone, are to be met with. Butchers give for such animals as much as £60 to £70 per head, while young steers between two and three years old make as much as £40 a head. Many are killed at an earlier age than this, great numbers being now slaughtered at two years old and under, which speaks volumes as to the early maturity to which the shorthorn attains. Still, it is very commonly thought that, while beef-making has been quite elevated to a science in the case of pure-bred shorthorns, the production of milk has been a good deal overlooked; and it stands to reason, notwithstanding the assertions of many to the contrary that shorthorn cows are good milkers, that this very tendency to put on flesh is opposed to the development and yielding of large quantities of milk, and for this purpose a cross with some other good milking breed is desirable where this object is required to be attained.

141. **FIFESHIRE AND LOWLAND SCOTCH.** — Fifeshire and Lowland Scotch cattle thrive on rich pastures and on good turnip

HEREFORD STEER.

soils, and will run with the Hereford and short-horned Yorkshire breeds and do very well.

142. **HEREFORDS.**—Herefords are supposed by many to be equal in value with the shorthorn as a breed, but they are not commonly met with far away from their native district the same as the more celebrated Durham breed are, and a distinction between the two has thus been drawn by Professor Low:—"The Herefords will frequently pay the graziers better than the Durhams; but the value of a breed is to be determined, not by the profit which it yields between buying and selling, but by that which it yields to the breeder and the feeder conjointly, from its birth to its maturity; and taking into account the early maturity of the shorthorns, and the weight to which they arrive, it may without error be asserted that they merit the preference which has been given to them."

I

143. THE GALLOWAY.—On poor land, affording but scanty herbage, the small Galloways and other Highland stock will be found to answer very well, as these do not lose their condition while there is only a short bite of grass in the summer; but these (not the very smallest kind), if kept on good pasture during the summer, and merely preserved from falling off in the winter, will attain a good size. It is stated in the "General Report of Scotland," upon the authority of well-known breeders, that these will weigh, at 2 to 2½ years of age, 30 stone; from 3 to 3½, 41 stone; and from 4 to 4½ years of age, 54 stone, the great addition to weight being acquired during the six months of the grass season. The average of our large-framed beasts in England of the best breeds, according to the testimony of carcase

SCOTCH POLLED BREED.

butchers and salesmen, is generally at 4 years old about 110 to 115 stone of 8 lbs. for the carcase, and 20 to 25 stone for the fat and hide, when they have been regularly grazed.

The quietness of the Galloway breed, and their readiness to fatten when their frame is in proper condition, causes them to be very favourably regarded, while, being without horns, a greater number can be kept together than is the case with horned cattle, which fight amongst themselves continually.

144. ABERDEEN, ANGUS, AND SMALL SCOTCH BREEDS.— There are various breeds of large cattle to be met with in the eastern districts of Scotland as well as Fifeshire, there being the Aberdeen and Angus breeds, some polled and some horned, being mostly black in colour; while in the extreme north, as in the Shetland Isles, an extremely small breed is to be met with, the cows of

which give a good deal of milk in proportion to their size; but the most noticeable feature in connection with this small breed is, that a cow crossed by a shorthorn bull will produce progeny which will attain an equal size with the larger breeds.

145. **THE AYRSHIRE.**—The Ayrshire bullocks are not found to answer well with the grazier. Their meat is coarse in quality, they are hard to fatten, and do not attain to great weights, but the cows, as before described, are capital milkers; and while, in England, great pains have been taken to improve the shorthorn for the sake of beef, in Scotland equal pains have been taken to develop the milking properties of the Ayrshire cows, which thrive and do well upon medium and even poor soils.

146. **THE LONGHORNS.**—The longhorns were a breed which at one time enjoyed considerable favour, especially in the Midland Counties of England, where they used universally to be met with; but they have given way to the shorthorn, which is more generally preferred, and they are now most commonly met with in some parts of Ireland, where they still retain their original reputation.

147. **THE ALDERNEY.**—The Alderney is the least valuable of any known breed for the grazier, though, as we have remarked before in another place, they weigh better in the scale than their appearance would warrant; but, as cows, they possess the merit of giving milk of superior excellence, while they do well on inferior pastures. Both Alderney and Ayrshire breeds are peculiarly butter and milk producing cows, the former being celebrated for the richness of the milk it yields, while the latter gives an unusually large quantity. It is often recommended that where Alderney cows are not regularly kept for milking purposes, one, at least, should be kept in a herd, as the admixture of her milk with the rest will sensibly improve the quality of the whole.

148. **WELSH CATTLE.**—Welsh cattle have generally a family likeness to the Highland breeds, but there are several distinct species. The Pembrokes thrive on poor soils, and the cows yield milk freely. They are alike useful animals to the cottager, whose opportunities for grazing them are but limited, and to gentlemen who may have grass of inferior quality to eat off in parks, or on mountain land. Anglesea are coarser and heavier than the Pembroke, while the Glamorgan, which are somewhat small, and inferior in those points looked for by the grazier, are yet good milkers, and the cows are appreciated on this score. They are seldom found out of the county from which they take their name, where the development

of the iron-works, and consequent increase of population, has
created a large demand for milk.

149. ADVICE ON BUYING.—From what we have already written
the intending buyer will be enabled to gather a good many practical
hints as to the class of animal most likely to be the best suited for
his own particular purpose and locality. It would be false economy
to buy cattle of unsuitable breeds merely because they may happen
to be cheap; for much more money would be wasted over injudicious
purchases, in the shape of the keep of an unprofitable animal, than
any likely to be gained in the form of a bargain secured in the
price below its apparent value.

It will always be found in the long run to answer the farmer's or grazier's purpose best to secure the exact breed of animals he considers most fitted for his
purpose—whatever that may be—and pay a fair price for them, rather than be
tempted by low rates to purchase cattle of an unsuitable description.

SECTION OF A COWHOUSE.

For grazing or feeding purposes some capital cattle are to be picked up out of
the droves of the West Highland cattle, or Kyloes. The true West Highland
ox has short muscular limbs, a wide and deep chest, finely-arched ribs, and
straight back, thick but mellow skin closely covered with shaggy hair. His head
is broad, with a short, fine muzzle, his eye full and bright, with long turned-up
horns, and a bold erect carriage, exhibiting, when of mature size and in good
condition, a symmetrical form and noble bearing which is difficult to be excelled
by any other breed. His compact carcase and the choice quality of beef cause
him to be a great favourite with butchers, while, contented with the coarsest
pasturage, he will ultimately fatten where shorthorns and similar breeds could
only manage with difficulty to keep life and body together.

150. ACCOMMODATION FOR CATTLE.—In arranging for the
accommodation of cattle it will be found of great assistance to have
a classified arrangement by which the cows of the dairy are kept
apart from the feeding stock, and the houses and yards for each
particular kind of stock kept as much as possible together, close to
which should be the sheds for storing and preparing the food.

In the case of a dairy farm, the out-houses should not be too far
from the farm-house, when domestic servants have a good deal of

labour to perform in them, such as milking, or carrying the milk to the dairy, and these should be the nearest, and the feeding cattle the farthest off.

The necessity for shelter, and the increased comfort to the animals and the improvement of their health from it, have been spoken of before. This should be supplied in accordance to the number of animals that are to be housed. If too great a space is allowed, it cannot be littered down properly, as too much will have to be traversed, which will prevent the litter from being properly converted into manure.

The house where food is consumed should be near to that where it is prepared, by which means a great deal of loss of time in unnecessary running about is saved.

Under the old system of shed and yard feeding the more valuable parts of the manure were exhaled into the atmosphere, or washed away by every shower, which ought to be caught and retained in a manure-tank.

The modern practice is to roof over the entire yard so as effectually to protect cattle, food, and manure from the vicissitudes of the weather, and to tie up the cattle for each meal and loosen them when they have eaten it, by which means they feed undisturbed, and yet get a certain amount of exercise.

The question has been much debated whether yards, stalls, or boxes are the best adapted for feeding cattle. Yards afford the greatest facilities for turning the straw into manure, but stalls require least litter, occupy the least space, and are more likely to be too warm than too cold, but deprive the animals of needful exercise, and they require more attendance.

Boxes combine, to a certain extent, the advantages of both these plans, as in them the animals are safe from cold and disturbance, get moderate exercise, require less attendance than those in stalls, and also less litter, while the manure made in them, being covered from the weather, retains the urine, and is superior on that account to manure made in open yards.

Warmth is one of the first essentials to fattening cattle. It is now well known that in the case of all warm-blooded animals a considerable portion of food is expended in maintaining the natural heat of their bodies, so that cattle exposed to a low temperature require an additional amount of food to keep up their necessary animal heat, which, if kept in by cover, will cause them to eat less, and yet lay on more fat.

151. LABOUR REQUIRED FOR SUPERINTENDING CATTLE. —The farmer will find it answer his purpose to give a good deal of personal superintendence to his cattle, as much depends upon the cribs being kept clean and the food regularly supplied only in the quantities that will be eaten.

Stale portions of food, or dirt left in the cribs, taints the fresh food, which is less relished, and, in consequence, does not do the animal so much good; and attention from some careful person, steadily persisted in, will amply repay the trouble that is taken.

Persons not accustomed to the management of cattle will find it of great advantage to spend a little time daily amongst them, and make their acquaintance by a little notice and occasional caress, as well as being thus able to identify each by their marks and general appearance.

A little familiarity of this kind accustoms them to the presence of persons, and they are not likely to be startled or give way to restless excitement when food and litter is supplied to them, or they are handled by strangers, possibly purchasers.

Sometimes the tying and untying has been objected to on the score of the extra labour involved; but it has been proved by repeated trials that two men can unloose a hundred cows in ten minutes and tie them up again in twenty minutes. The herd-boy who waits on the cows in the field stands at the door to prevent too many crushing in at one time.

The unloosing of stock is often found of advantage, and when there are not boxes and the buildings have to be made use of already established, where the cattle are placed in yards with sheds around them for shelter, the experience of graziers has shown that the beasts will often eat food thrown to them on the ground which they will reject when offered to them in the stalls. Although at first the operation of tying and untying may give a good deal of trouble, practice makes it very easy eventually, and the beasts are benefited by being loosened for a short time when the system of tying-up is followed; and they should be put up together as much as possible of the same age and strength; if not, the strong will prevent the weak from feeding until they themselves are satisfied.

Cattle that have been reared together can be packed closer than those which have been bought from dealers and collected promiscuously, six being the average number which should be fed together when the size and construction of the sheds permit it.

152. FEEDING.—The system of feeding cattle hitherto has been chiefly to allow them as many sliced turnips as they could consume, and the racks supplied with fresh oat straw daily. Straw, as an article of food, has been in the past very much wasted. The digestive organs of the ox are formed with a manifest adaptation to the consumption of very bulky and but moderately-nutritious food, such as grass or hay, and he must have his fill before he composes himself to rest and commences to ruminate.

By being allowed to eat a large quantity of richer food, not only is a greater expense incurred, but as the animal's powers of assimilation are not equal to its proper digestion, the wasted surplus produces irritation and disturbance, which is often made plainly apparent by continued diarrhœa, and sometimes by more serious disease.

It is necessary that his capacious paunch be constantly full, and straw can be made to play a very prominent part in this proceeding.

153. **STRAW AS FOOD FOR CATTLE.**—Mr. Joseph Darby has pointed out in a very useful pamphlet—a reprint from a paper which appeared in the Journal of the Royal Agricultural Society—the great advantages that accrue from using straw as food for stock, The nutritive qualities of straw are very various, and differ with its stages of ripeness, which will be referred to in the quotations which follow.

Mr. Darby says:—"As the results of chemical analysis, Dr. Voelcker has placed the nutritive values of different sorts of straw in the following order:— 1. Pea-straw. 2. Oat-straw. 3. Bean-straw with the pods. 4. Barley-straw. 5. Wheat-straw. 6. Bean-straw without the pods. The testimony of practical farmers has pretty generally endorsed this qualification. Pea-straw has always been considered too valuable to be used as litter, and it generally falls to the lot of sheep, these animals being particularly fond of it. Nearly all my correspondents set a higher value on oat-straw than on any other white straw for feeding purposes. There is less unanimity with regard to the virtue of barley-straw, attributable, no doubt, to the fact that its feeding value is not unfrequently materially increased by the large quantities of young clover mown with it. When there is little of this it very often sinks below wheat-straw in the scale of value, owing to the usual and almost invariable practice of over-ripening the barley crop. The custom of doing this cannot, of course, be urged against, as the grain is improved thereby for malting purposes; but both corn and straw of wheat would no doubt be improved if farmers could only more generally be induced to take the crop from the ground somewhat earlier than they are accustomed to do at present. No kind of straw probably differs more materially in value than that of the bean crop; and some admits of being heightened in quality by the beans being either cut or pulled while the stalks are green, and before the leaves have all dropped off. When beans grow to the height of seven or eight feet, as I have sometimes seen them, the stalks, of course, are like sticks; and should the crop be allowed to get dead ripe, it would be very ill-adapted to yield food without being chaffed and steamed. But if the Russian, or winter bean, be cultivated, which is short in the haulm and ripens in July, and if the crop be taken from the soil early while yet green, an exceedingly valuable straw for foddering or chaffing purposes would naturally be the result. One of the best farmers in South Hants used to be very fond of having his winter beans pulled up in that condition and placed in rows of stooks after being sheaved. This allowed the land to be cropped with turnips, and I have often heard him declare that, while he obtained a fairer sample of grain, worth several shillings a quarter more than ordinary samples, the bean-straw was also rendered of great value in affording material for utilisation as food for stock.

"Who can doubt that when farmers find it to their interest to care more than they now do about straw produce, so as to secure it in a condition better adapted to serve for fodder, similar tactics will be employed in harvesting all crops, with the exception, perhaps, of barley? Nothing more surely need be stated as to the advantages of cutting oats early; and yet there is another point materially bearing on the matter which has not yet been mentioned. Oat-corns adhere to the plant by so frail a thread, that if the crop be allowed thoroughly to ripen, large numbers of them are tolerably sure to bret out by the first strong wind which blows. Every experienced man knows how hazardous it is to allow oats to remain uncut after the straw begins to turn off in colour. There is, consequently, every inducement to harvest that crop early. When also it is considered to what an extent both the grain and the straw of wheat are improved by the cutting being effected just at the period when the corns no longer emit a milky juice, common sense naturally points to the proper course of action. All these things vitally affect the issue, and we shall perhaps soon find even the occupiers of the Fens and our richest alluvial soils ready to admit that, by altering their course of action slightly, in taking their grain crops from the ground earlier than they have hitherto done, a great deal more may be made out of straw. The farmers of Lincolnshire, who, by growing green crops bulky and coarse in straw, fancy there is little feeding virtue in it, are still accustomed even now to utilize no small portion as food, by their stock being allowed to pick out of large quantities the tit-bits and stalk tops. By adopting earlier cutting they would, no doubt, find a means of economic management hitherto only partially explored.

" Nor must it be forgotten that, however much the coarseness of texture and the condition of the straw in different districts may affect their value for feeding purposes, the best of the best would not be worth much given singly, without the addition of rich substances, such as oil-cake or corn-meal, with root-pulp or roots, should the latter be plentiful. Only as an ingredient in a mixed dietary for stock can straw yield fully the advantages it is capable of rendering as a food substance. This does not imply that straw should be utilized in this way or that. Many farmers like to save expense; and it is natural, perhaps, that the material, when exceedingly abundant, and not of the finest texture or quality, should be given whole and in large quantity; but still, if the animals are at the same time fed with sufficient liberality on richer substances, so as to keep them laying on flesh actively, or yielding milk bountifully, or, if young, in active growth and thriving condition, the principal object will be attained.

" However strange to the ears of some it may sound to hear of beef, mutton, or butter being derived as the direct result of feeding on straw, this appears to be the most economical way of producing either of those high-priced articles in winter, provided that straw forms one item only in the dietary, of which the other items should be roots and oil-cake, or corn, as a rule, but varied with other rich and suitable ingredients if they be cheaper to purchase, or more adapted to the wants of the animals."

Mr. Mechi says :—"If we are to consume all our bean, barley, wheat, and oat-straw, we must keep our animals on sparred floors, or on burnt clay, and we must invest more capital in animals. We shall then make much more meat per acre. If a ton of straw will make 40 lbs. of meat, and if two tons of straw are grown per acre of our cereal and pulse crops, it would be four-score pounds of meat per acre over the whole of the cereals and pulse."

Of course Mr. Mechi did not mean that it is possible to make so much meat out of the straw, unless it be given in conjunction with auxiliary feeding stuffs: for immediately afterwards, to quiet any apprehensions as to the manure-heap being lessened in value, he says, " Your animal, by this mode of feeding, consumes 560 lbs. of rape-cake with every ton of straw." Dr. Voelcker, and other scientific experts, have, I believe, sufficiently proved by chemical analysis that a ton of straw possesses sufficient nutritive properties to yield this amount of beef; but an animal could not eat enough straw to keep the machinery going without the addition of richer feeding substances. The whole virtue would be taken up in supplying heat to the system, and repairing the waste of the tissues, &c. But when straw is used for bulk, and oil-cake and other substances to improve the quality of a mixed dietary, it is only reasonable to give the straw credit for what it supplies towards the beef-making; and this appears to be what Mr. Mechi has actually done. Mr. Horsfall, in the Journal, observed: "In wheat-straw, for which I pay 35*s.* per ton, I obtain for 1*s.* 2½*d.* '50 oil and 32 lbs. of starch, or (the starch reduced to oil) 18 lbs. available for the production of fat or for respiration. I know no other material from which I can derive by purchase an equal amount of this element of food at so low a price. The value of straw calculated as manure is 9*s.* 7*d.* per ton."

But Mr. Horsfall gave this as scientific evidence, fully accounting for his success in a particular system of feeding dairy cows on a mixed dietary, the chief items of which were rape-cake, malt-combs, bran, and straw-chaff of different kinds, all intermixed and steamed, or cooked before being employed. The results were so important, that his cows gave more bountiful yieldings of milk, and of far higher quality than they had done before, and put on flesh rapidly, even to getting quite fat, while in full profit. His cream was of so thick a consistency as to admit of laying a penny piece on it without sinking, and it yielded a far larger proportion of butter than ordinary cream. Casting about for reasons to account for all this, he found them in a comparison of the chemical analysis of the mixed nutritive substances supplied by him, with that of the food commonly supplied to dairy cows. His researches led him to see that even the best hay is not a food good enough for a milch-cow to enable her to do her best; and, he said, "You cannot induce a cow to consume the quantity of hay requisite for her maintenance, and for a full yield of milk."

Mr. Horsfall fully proved, both scientifically and practically, the greater economy of feeding milch-cows on straw-chaff, rape-cake, malt-combs, &c., rather than on hay; but the immense value of straw to him consisted in his system allowing the full amount of nutritive properties it contains to be appropriated. That this was his own view appears from the following:—"I am satisfied the most economical use of food rich in albuminous matter is together with straw and other materials which are deficient in this element."

154. SUMMER FEEDING.—In feeding cattle upon the summer soiling system of giving green stuff, care should be taken not to give too much in the first place, as the greediness of the animals after having long been kept upon dry food causes the accident we have previously described as "hoving," when the gases arising from tares, clover, lucerne, &c., cause swelling of the stomach, which obstructs rumination, and sometimes even causes death. This may be prevented by the use of straw, which also corrects the tendency to looseness of the bowels, which is apt to arise from too free use of green food; and it will be found by far the best plan not to make too sudden changes in this respect, but to accustom the animals gradually to the change of food which the annual recurrence of the seasons brings round. Thus, instead of giving them all green food at once, these grasses should be mixed with chopped straw, and by a like system of management, when the green stuff gets scarce the way for the drier food should be prepared in the same considerate manner.

155. **WINTER FEEDING.**—In winter feeding the lavish quantity of roots often used can be reduced with positive advantage. It has been proved that a medium-sized bullock will improve faster when only 80 or 100 lbs. of turnips are given to it daily, with straw, than when allowed to eat 2 cwt. of turnips, which he will do if he gets the chance. The difficulty in getting cattle to eat straw in sufficient quantity can be obviated by reducing it to chaff by means of a straw-cutter, and mixing with it small quantities of bruised linseed, bean, or other meal, and by infusion in boiling water, or steaming in a close vessel, so incorporating the ingredients that a grateful flavour is imparted to the straw, and a willing consumption of this bulky factor is induced.

Mr. Warnes, of Trimmingham, relates his plan of feeding with linseed as follows:—"I commenced winter-feeding this year upon white turnips grown after flax, the tops of which, being very luxuriant, are cut with pea-straw into chaff, compounded with linseed-meal, and given to my bullocks according to the following plan:—Upon every six pails of boiling water, one of finely-crushed linseed-meal is sprinkled by the hand of one person, while another rapidly stirs it round. (The advantage of this plan may be seen in the superior quality which results from making porridge in this way, with which children are fed, over that where merely the barley-meal is flung into the pot or saucepan at once, when a marked difference in quality is apparent.) In five minutes, the mucilage being formed, a half-hogshead is placed close to the boiler, and a bushel of the cut turnip tops and straw put in. Two or three hand-cupfuls of the mucilage are then poured upon it, and stirred in with a common muck-fork. Another bushel of the turnip-tops, chaff, &c., is next added, and two or three cups of the jelly as before; all of which is then expeditiously stirred and worked together with the fork and rammer. It is afterwards pressed down as firmly as the nature of the mixture will allow with the latter instrument, which completes the first layer. Another bushel of the pea-straw, chaff, &c., is thrown into the tub, the mucilage poured upon it as before, and so on till the boiler is emptied. The contents of the tub are lastly smoothed over with a trowel, covered down, and in two or three hours the straw, having absorbed the mucilage, will also with the turnip-tops have become partially cooked. The compound is then usually given to the cattle, but sometimes is allowed to remain till cold. The bullocks, however, prefer it warm; but whether hot or cold, devour it with avidity."

Mr. Ogden, Berry Hill, Northumberland, in a report read to the East of Berwick Farmers' Club, described his plan of feeding as follows:—"My cattle are fed with turnips, bean-meal, oil-cake, and cut straw. The first thing in the morning they get the mixture, then turnips, and at one o'clock the mixture again; afterwards turnips. On Sundays the mixture is withheld. I find that a three-year-old steer will consume (if fed on them alone) from 16 to 18 stones of turnips daily. The mixture I am in the habit of giving to my cattle consists of 2 lbs. of oil-cake, 2 lbs. of bean-meal, 4 lbs. of cut straw, and 1½ oz. of salt daily. This mixture can be purchased and prepared, at present prices, for 1d. per pound, or 2s. per head per week, six days in the week. I also find that cattle, when they have this mixture, consume at least 1 cwt. of turnips *less* per day than when fed upon turnips alone. This mixture is prepared in the forenoon by the byre-man, and keeps perfectly sweet for thirty-six hours. In preparing the mixture, to serve 24 cattle for 24 hours, 48 lbs. of oil-cake, 48 lbs. of bean-meal, 96 lbs. cut straw, and 30 oz. of salt are, in the first place, well mixed together in a trough; 36 gallons of boiling water are then added, after which the whole mass is well turned and incorporated together and pressed down; and in an hour or two is quite ready for the cattle. The troughs in which this mixture is prepared are 6 feet long, 2 feet wide, and 2½ feet deep. A trough of this size will contain mixture for twenty-four cattle, and the time occupied by the byre-man in preparing one trough-full of the mixture is not more than half an hour, the cut straw, meal. &c., being all ready."

The advantage of giving steamed food to cattle is very great. Straw that has been threshed for some time loses its freshness, and even hay is often a little mouldy, so that it is no longer welcome to the cattle. All this, however, is disposed of in the course of steaming and mixing with meal and other rich substances, so that hay or straw which has been rejected is eaten readily when accompanied with other appetising ingredients.

A great many examples of different methods adapted for feeding stock profitably have been instanced under another heading, so we shall here only briefly remark that, when grasses and dry food are mixed, it will be found best to make the mixture overnight when the dry provender will be found to have acquired a sweet vegetable taste, to which we have previously referred, which the animals relish exceedingly.

Some farmers attempt to feed cattle upon straw by itself when they are hungry, before giving them the more inviting food, but in course of time they will come to reject it; but no ill-consequences from irregular feeding and eating can arise if the food is carefully and properly mixed beforehand. It is the want of the necessary pains and precautions which ought to be taken in feeding stock that causes it to be less profitable very often than it otherwise would be. The fatting cattle, when turnips are given, should have the bulbs, and the green tops and top roots should be given to the store stock.

Experiments have shown that different breeds of cattle will acquire various degrees of substance or flesh from the same quantity of food supplied to each; and these and similar points deserve careful attention and notice, for no rule can be drawn as to the exact quantity of food required by each beast.

156. CATTLE FARMING ABROAD.—It is clearly apparent that, in economical feeding, a large portion of the profit is to be found where a number of animals are kept; and in Germany, especially, many economical contrivances are resorted to for eking out the food of stock that are not practised in England. We have touched upon this subject under a distinct heading. In the district of the Lower Moselle, as we pointed out, in the spring, the women and children range the fields, and cut the young thistles and nettles,

MONTAFUN COW.

EGERLAND COW.

Murzthal Cow.

Pinzgau Cow.

and dig up the roots of the couch-grass, collect weeds of all kinds, and strive to turn them to account.

What is thus scraped together is well washed, mixed with cut straw and chaff, and after boiling water has been poured over the whole, it is given to the cattle, which are stall fed.

On the other hand, in Moldavia and Bessarabia, the cattle are kept in the fields all the year round, exposed to all the inclemencies of the weather.

157. **FOREIGN BREEDS.**—Our business is mainly with the ordi-

PODOLIAN COW.

nary stock that is commonly found in the United Kingdom, but the following illustrations of Austrian cattle will doubtless be considered interesting to many.

The Podolian is an aboriginal race of cattle distinguished by its capability of enduring changes of weather, and contentedness with poor fare.

The Murzthal breed is appreciated on account of its milking properties, and as draught oxen.

The Montafun are distinguished for good temper, and belong to the heavy average group of cattle.

The Egerland are noticeable on account of their general healthiness, and contentedness with the quality and amount of food given to them.

The Pinzgau is a breed that is distributed throughout the whole of the Salzburg region.

The Kuhland cattle, though only of the middle height, must yet be classed with the heavier races of stock.

KUHLAND COW.

LONDON:

PRINTED BY J. OGDEN AND CO.,
172, ST. JOHN STREET, E.C.

INDEX.

K

A SELECTION

FROM

WARD, LOCK & CO.'S

CATALOGUE

OF

𝔑𝔢𝔴 𝔞𝔫𝔡 𝔓𝔬𝔭𝔲𝔩𝔞𝔯 𝔅𝔬𝔬𝔨𝔰.

Of all Works of Reference published of late years, not one has gained such general approbation as BEETON'S ILLUSTRATED ENCYCLOPÆDIA. *The importance of this valuable compilation in the cause of mental culture has long been acknowledged, and of its real usefulness to the public, the most gratifying proofs have been received. It is undoubtedly one of the Most Comprehensive Works in existence, and is*

THE CHEAPEST ENCYCLOPÆDIA EVER PUBLISHED.

Complete in Four Volumes, demy 8vo, half-roan, price 42*s*.

BEETON'S

ILLUSTRATED ENCYCLOPÆDIA

OF UNIVERSAL INFORMATION.

COMPRISING

GEOGRAPHY, HISTORY, BIOGRAPHY ART, SCIENCE, AND LITERATURE,

AND CONTAINING

4000 Pages, 50,000 Articles, and 2,000 Engravings and Coloured Maps.

In BEETON'S ILLUSTRATED ENCYCLOPÆDIA will be found complete and authentic information respecting the Physical and Political Geography, Situation, Population, Commerce and Productions, as well as the principal Public Buildings, of every Country and important or interesting Town in the World, and the leading Historical Events with which they have been connected ; concise Biographies of Eminent Persons, from the most remote times to the present day ; brief Sketches of the leading features of Egyptian, Greek, Roman, Oriental, and Scandinavian Mythology ; a Complete Summary of the Moral, Mathematical, Physical and Natural Sciences ; a plain description of the Arts ; and an interesting Synopsis of Literary Knowledge. The Pronunciation and Etymology of every leading term introduced throughout the Encyclopædia are also given.

London : *WARD, LOCK & CO., Salisbury Square, E.C.*

THE STANDARD COOKERY BOOKS.

s.	d.	
3	6	**MRS. BEETON'S EVERY-DAY COOKERY AND** HOUSEKEEPING BOOK. Comprising Instructions for Mistresses and Servants, and a Collection of over 1,650 Practical Recipes. With Hundreds of Engravings in the Text, and 142 Coloured Figures showing the Modern Mode of sending Dishes to Table. Cloth gilt, price 3s. 6d.
2	6	**MRS. BEETON'S ALL ABOUT COOKERY.** A Collection of Practical Recipes, arranged in Alphabetical Order, and
2	0	fully Illustrated. Crown 8vo, cloth plain, price 2s. ; or in new and improved binding, cloth gilt, price 2s. 6d.
1	6	**MRS. BEETON'S ENGLISHWOMAN'S COOKERY** BOOK. An Entirely New Edition, Revised and Enlarged. Containing upwards of 600 Recipes, 100 Engravings and Four Coloured Plates. With Directions for Marketing, Diagrams of Joints, Instruc-
1	0	tions for Carving, the Method of folding Table Napkins, &c., and Descriptions of Quantities, Times, Costs, Seasons, for the various Dishes. Post 8vo, cloth, price 1s. ; cloth gilt, half bound, price 1s. 6d.
1	0	**THE ARTIZANS' AND WORKPEOPLE'S COOK-** ERY BOOK. Containing a large number of Economical Recipes, together with Explanations as to the principles of Cookery—Food ; What it Is, and what it Does—What to Eat and When—Things in Season— How to Keep Food—Kitchen Utensils, &c. Price 1s.
1	0	**THE PEOPLE'S HOUSEKEEPER.** A Complete Guide to Comfort, Economy, and Health. Comprising Cookery, Household Economy, the Family Health, Furnishing, Housework, Clothes, Marketing, Food, &c., &c. Post 8vo, cloth, price 1s.
0	6	**THE SIXPENNY PRACTICAL COOKERY AND** ECONOMICAL RECIPES. Comprising Marketing, Relishes, Boiled Dishes, Vegetables, Soups, Side Dishes, Salads, Stews, Fish, Joints, Sauces, Cheap Dishes, Invalid Cookery, &c. Price 6d.
0	6	**THE SIXPENNY ECONOMICAL COOKERY** BOOK, for Housewives, Cooks, and Maids-of-all-Work ; with Advice to Mistress and Servant—Gossip to Young Mistresses—Dinners— Poultry and Game—Fish—Pastry and Puddings—Vegetables—Preserves—Sick Room Diet—Useful Sundries—Hints for Comfort and Cleanliness. By Mrs. WARREN. Post 8vo, linen covers, price 6d.
0	6	**THE COOKERY LESSON BOOK,** for Schools and Young Housekeepers. An Easy and Complete Guide to Economy in the Kitchen. A valuable Handbook for Young Housewives. Price 6d.
0	1	**BEETON'S PENNY COOKERY BOOK.** Entirely New Edition, with New Recipes throughout. Three Hundred and Thirtieth Thousand. Containing more than Two Hundred Recipes and Instructions. Price 1d. ; post free, 1½d.
0	1	**WARD AND LOCK'S PENNY HOUSEKEEPER** AND GUIDE TO COOKERY. Containing Plain and Reliable Instructions in Cleaning and all Domestic Duties, the Preparation of Soups, Vegetables, Meats of all kinds, Pastry, Jellies, Bread, Home Beverages, &c., and everything necessary for securing a well-ordered Home. Price 1d. ; post free, 1½d.
0	1	**BEETON'S PENNY DOMESTIC RECIPE BOOK :** Containing Simple and Practical Information upon things in general use and necessary for every Household. Price 1d.; post free, 1½d.

London : WARD, LOCK & CO., *Salisbury Square, E.C.*

THE PEOPLE'S STANDARD CYCLOPÆDIAS.

EVERYBODY'S LAWYER (Beeton's Law Book). Entirely
New Edition, Revised by a BARRISTER. A Practical Compendium of the
General Principles of English Jurisprudence: comprising upwards of 14,600
Statements of the Law. With a full Index, 27,000 References, every numbered
paragraph in its particular place, and under its general head. Crown 8vo, 1,680
pp., cloth gilt, 7s. 6d.

** The sound practical information contained in this voluminous work is
equal to that in a whole library of ordinary legal books, costing many guineas.
Not only for every non-professional man in a difficulty are its contents valuable,
but also for the ordinary reader, to whom a knowledge of the law is more important and interesting than is generally supposed.

BEETON'S DICTIONARY OF GEOGRAPHY: A Universal
Gazetteer. Illustrated by Maps—Ancient, Modern, and Biblical, and several
Hundred Engravings in separate Plates on toned paper. Containing upwards
of 12,000 distinct and complete Articles. Post 8vo, cloth gilt, 7s. 6d.; half-calf,
10s. 6d.

BEETON'S DICTIONARY OF BIOGRAPHY: Being the
Lives of Eminent Persons of All Times. Containing upwards of 10,000 distinct
and complete Articles, profusely Illustrated by Portraits. With the Pronunciation of Every Name. Post 8vo, cloth gilt, 7s. 6d.; half-calf, 10s. 6d.

BEETON'S DICTIONARY OF NATURAL HISTORY:
A Popular and Scientific Account of Animated Creation. Containing upwards
of 2,000 distinct and complete Articles, and more than 400 Engravings. With
the Pronunciation of Every Name. Crown 8vo, cloth gilt, 7s. 6d.; half-calf,
10s. 6d.

BEETON'S BOOK OF HOME PETS: How to Rear and
Manage in Sickness and in Health. With many Coloured Plates, and upwards
of 200 Woodcuts from designs principally by HARRISON WEIR. With a Chapter
on Ferns. Post 8vo, half-bound, 7s. 6d.

THE TREASURY OF SCIENCE, Natural and Physical.
Comprising Natural Philosophy, Astronomy, Chemistry, Geology, Mineralogy,
Botany, Zoology and Physiology. By F. SCHOEDLER, Ph.D. Translated and
Edited by HENRY MEDLOCK, Ph.D., &c. With more than 500 Illustrations.
Crown 8vo, cloth gilt, 7s. 6d.

A MILLION OF FACTS of Correct Data and Elementary In-
formation concerning the entire Circle of the Sciences, and on all subjects of
Speculation and Practice. By Sir RICHARD PHILLIPS. Carefully Revised and
Improved. Crown 8vo, cloth gilt, 7s. 6d.

THE TEACHER'S PICTORIAL BIBLE AND BIBLE
DICTIONARY. With the most approved Marginal References, and Ex-
planatory Oriental and Scriptural Notes, Original Comments, and Selections
from the most esteemed Writers. Illustrated with numerous Engravings and
Coloured Maps. Crown 8vo, cloth gilt, red edges, 8s. 6d.; French morocco,
10s. 6d.; half-calf, 10s. 6d.

THE SELF-AID CYCLOPÆDIA, for Self-Taught Students.
Comprising General Drawing; Architectural, Mechanical, and Engineering
Drawing; Ornamental Drawing and Design; Mechanics and Mechanism; the
Steam Engine. By ROBERT SCOTT BURN, F.S.A.E., &c. With upwards of
1,000 Engravings. Demy 8vo, half-leather, price 10s. 6d.

London: WARD, LOCK & CO., Salisbury Square, E.C.

REFERENCE BOOKS FOR THE PEOPLE.

BEETON'S NATIONAL REFERENCE BOOKS,

FOR THE PEOPLE OF GREAT BRITAIN AND IRELAND.

Each Volume Complete in itself, and containing from 512 to 590 Columns.

** *In an age of great competition and little leisure the value of Time is tolerably well understood. Men wanting facts like to get at them with as little expenditure as possible of money or minutes.* BEETON'S NATIONAL REFERENCE BOOKS *have been conceived and carried out in the belief that a set of Cheap and Handy Volumes in Biography, Geography, History (Sacred and Profane), Science, and Business, would be thoroughly welcome, because they would quickly answer many a question. In every case the type will be found clear and plain.*

STRONGLY BOUND IN CLOTH, PRICE ONE SHILLING EACH; or cloth gilt, 1s. 6d.

1. Beeton's British Gazetteer: A Topographical and Historical Guide to the United Kingdom.

2. Beeton's British Biography: From the Earliest Times to the Accession of George III.

3. Beeton's Modern Men and Women: A British Biography, from the Accession of George III. to the Present Time.

4. Beeton's Bible Dictionary. A Cyclopædia of the Geography, Biography, Narratives, and Truths of Scripture.

5. Beeton's Classical Dictionary : A Cyclopædia of Greek and Roman Biography, Geography, Mythology, and Antiquities.

6. Beeton's Medical Dictionary. A Guide for every Family, defining, with perfect plainness, the Symptoms and Treatment of all Ailments, Illnesses, and Diseases.

7. Beeton's Date Book. A British Chronology from the Earliest Records to the Present Day.

8. Beeton's Dictionary of Commerce. Containing Explanations of the principal Terms used in, and modes of transacting Business at Home and Abroad.

9. Beeton's Modern European Celebrities. A Biography of Continental Men and Women of Note who have lived during the last Hundred Years, or are now living.

Beeton's Guide Book to the Stock Exchange and Money Market. With Hints to Investors and the Chances of Speculators. Entirely New Edition, post 8vo, linen covers, 1s.

Beeton's Investing Money with Safety and Profit. New and Revised Edition. Post 8vo, linen covers, 1s.

Beeton's Ready Reckoner. With New Tables, and much Information never before collected. Post 8vo, strong cloth, 1s.

Webster's Sixpenny Ready Reckoner. Demy 32mo, 256 pp., strong cloth, 6d.

Beeton's Complete Letter Writer, for Ladies and Gentlemen. Post 8vo, strong cloth, price 1s.

Beeton's Complete Letter Writer for Ladies. In linen covers,6d.

Beeton's Complete Letter Writer for Gentlemen. Ditto.

The New Letter Writer for Lovers. Ditto.

Webster's Shilling Book-keeping. A Comprehensive Guide, comprising a Course of Practice in Single and Double Entry. Post 8vo, cloth, 1s.

London : WARD, LOCK & CO., Salisbury Square, E.C.

THE
COW

Perkins

Ingram Content Group UK Ltd.
Milton Keynes UK
UKHW020918140323
418553UK00007B/530